Computers in Geometry and Topology

LECTURE NOTES

IN PURE AND APPLIED MATHEMATICS

1. *N. Jacobson,* Exceptional Lie Algebras
2. *L. -Å. Lindahl and F. Poulsen,* Thin Sets in Harmonic Analysis
3. *I. Satake,* Classification Theory of Semi-Simple Algebraic Groups
4. *F. Hirzebruch, W. D. Newmann, and S. S. Koh,* Differentiable Manifolds and Quadratic Forms (out of print)
5. *I. Chavel,* Riemannian Symmetric Spaces of Rank One (out of print)
6. *R. B. Burckel,* Characterization of C(X) Among Its Subalgebras
7. *B. R. McDonald, A. R. Magid, and K. C. Smith,* Ring Theory: Proceedings of the Oklahoma Conference
8. *Y.-T. Siu,* Techniques of Extension on Analytic Objects
9. *S. R. Caradus, W. E. Pfaffenberger, and B. Yood,* Calkin Algebras and Algebras of Operators on Banach Spaces
10. *E. O. Roxin, P.-T. Liu, and R. L. Sternberg,* Differential Games and Control Theory
11. *M. Orzech and C. Small,* The Brauer Group of Commutative Rings
12. *S. Thomeier,* Topology and Its Applications
13. *J. M. Lopez and K. A. Ross,* Sidon Sets
14. *W. W. Comfort and S. Negrepontis,* Continuous Pseudometrics
15. *K. McKennon and J. M. Robertson,* Locally Convex Spaces
16. *M. Carmeli and S. Malin,* Representations of the Rotation and Lorentz Groups: An Introduction
17. *G. B. Seligman,* Rational Methods in Lie Algebras
18. *D. G. de Figueiredo,* Functional Analysis: Proceedings of the Brazilian Mathematical Society Symposium
19. *L. Cesari, R. Kannan, and J. D. Schuur,* Nonlinear Functional Analysis and Differential Equations: Proceedings of the Michigan State University Conference
20. *J. J. Schäffer,* Geometry of Spheres in Normed Spaces
21. *K. Yano and M. Kon,* Anti-Invariant Submanifolds
22. *W. V. Vasconcelos,* The Rings of Dimension Two
23. *R. E. Chandler,* Hausdorff Compactifications
24. *S. P. Franklin and B. V. S. Thomas,* Topology: Proceedings of the Memphis State University Conference
25. *S. K. Jain,* Ring Theory: Proceedings of the Ohio University Conference
26. *B. R. McDonald and R. A. Morris,* Ring Theory II: Proceedings of the Second Oklahoma Conference
27. *R. B. Mura and A. Rhemtulla,* Orderable Groups
28. *J. R. Graef,* Stability of Dynamical Systems: Theory and Applications
29. *H.-C. Wang,* Homogeneous Branch Algebras
30. *E. O. Roxin, P.-T. Liu, and R. L. Sternberg,* Differential Games and Control Theory II
31. *R. D. Porter,* Introduction to Fibre Bundles
32. *M. Altman,* Contractors and Contractor Directions Theory and Applications
33. *J. S. Golan,* Decomposition and Dimension in Module Categories
34. *G. Fairweather,* Finite Element Galerkin Methods for Differential Equations
35. *J. D. Sally,* Numbers of Generators of Ideals in Local Rings
36. *S. S. Miller,* Complex Analysis: Proceedings of the S.U.N.Y. Brockport Conference
37. *R. Gordon,* Representation Theory of Algebras: Proceedings of the Philadelphia Conference
38. *M. Goto and F. D. Grosshans,* Semisimple Lie Algebras
39. *A. I. Arruda, N. C. A. da Costa, and R. Chuaqui,* Mathematical Logic: Proceedings of the First Brazilian Conference

40. *F. Van Oystaeyen,* Ring Theory: Proceedings of the 1977 Antwerp Conference
41. *F. Van Oystaeyen, and A. Verschoren,* Reflectors and Localization: Application to Sheaf Theory
42. *M. Satyanarayana,* Positively Ordered Semigroups
43. *D. L. Russell,* Mathematics of Finite-Dimensional Control Systems
44. *P.-T. Liu and E. Roxin,* Differential Games and Control Theory III: Proceedings of the Third Kingston Conference, Part A
45. *A. Geramita and J. Seberry,* Orthogonal Designs: Quadratic Forms and Hadamard Matrices
46. *J. Cigler, V. Losert, and P. Michor,* Banach Modules and Functors on Categories of Banach Spaces
47. *P.-T. Liu and J. G. Sutinen,* Control Theory in Mathematical Economics: Proceedings of the Third Kingston Conference, Part B
48. *C. Byrnes,* Partial Differential Equations and Geometry
49. *G. Klambauer,* Problems and Propositions in Analysis
50. *J. Knopfmacher,* Analytic Arithmetic of Algebraic Function Fields
51. *F. Van Oystaeyen,* Ring Theory: Proceedings of the 1978 Antwerp Conference
52. *B. Kedem,* Binary Time Series
53. *J. Barros-Neto and R. A. Artino,* Hypoelliptic Boundary-Value Problems
54. *R. L. Sternberg, A. J. Kalinowski, and J. S. Papadakis,* Nonlinear Partial Differential Equations in Engineering and Applied Science
55. *B. R. McDonald,* Ring Theory and Algebra III: Proceedings of the Third Oklahoma Conference
56. *J. S. Golan,* Structure Sheaves over a Noncommutative Ring
57. *T. V. Narayana, J. G. Williams, and R. M. Mathsen,* Combinatorics, Representation Theory and Statistical Methods in Groups: YOUNG DAY Proceedings
58. *T. A. Burton,* Modeling and Differential Equations in Biology
59. *K. H. Kim and F. W. Roush,* Introduction to Mathematical Consensus Theory
60. *J. Banas and K. Goebel,* Measures of Noncompactness in Banach Spaces
61. *O. A. Nielson,* Direct Integral Theory
62. *J. E. Smith, G. O. Kenny, and R. N. Ball,* Ordered Groups: Proceedings of the Boise State Conference
63. *J. Cronin,* Mathematics of Cell Electrophysiology
64. *J. W. Brewer,* Power Series Over Commutative Rings
65. *P. K. Kamthan and M. Gupta,* Sequence Spaces and Series
66. *T. G. McLaughlin,* Regressive Sets and the Theory of Isols
67. *T. L. Herdman, S. M. Rankin, III, and H. W. Stech,* Integral and Functional Differential Equations
68. *R. Draper,* Commutative Algebra: Analytic Methods
69. *W. G. McKay and J. Patera,* Tables of Dimensions, Indices, and Branching Rules for Representations of Simple Lie Algebras
70. *R. L. Devaney and Z. H. Nitecki,* Classical Mechanics and Dynamical Systems
71. *J. Van Geel,* Places and Valuations in Noncommutative Ring Theory
72. *C. Faith,* Injective Modules and Injective Quotient Rings
73. *A. Fiacco,* Mathematical Programming with Data Perturbations I
74. *P. Schultz, C. Praeger, and R. Sullivan,* Algebraic Structures and Applications Proceedings of the First Western Australian Conference on Algebra
75. *L. Bican, T. Kepka, and P. Nemec,* Rings, Modules, and Preradicals
76. *D. C. Kay and M. Breen,* Convexity and Related Combinatorial Geometry: Proceedings of the Second University of Oklahoma Conference
77. *P. Fletcher and W. F. Lindgren,* Quasi-Uniform Spaces
78. *C.-C. Yang,* Factorization Theory of Meromorphic Functions
79. *O. Taussky,* Ternary Quadratic Forms and Norms
80. *S. P. Singh and J. H. Burry,* Nonlinear Analysis and Applications
81. *K. B. Hannsgen, T. L. Herdman, H. W. Stech, and R. L. Wheeler,* Volterra and Functional Differential Equations

82. *N. L. Johnson, M. J. Kallaher, and C. T. Long,* Finite Geometries: Proceedings of a Conference in Honor of T. G. Ostrom
83. *G. I. Zapata,* Functional Analysis, Holomorphy, and Approximation Theory
84. *S. Greco and G. Valla,* Commutative Algebra: Proceedings of the Trento Conference
85. *A. V. Fiacco,* Mathematical Programming with Data Perturbations II
86. *J.-B. Hiriart-Urruty, W. Oettli, and J. Stoer,* Optimization: Theory and Algorithms
87. *A. Figa Talamanca and M. A. Picardello,* Harmonic Analysis on Free Groups
88. *M. Harada,* Factor Categories with Applications to Direct Decomposition of Modules
89. *V. I. Istrăţescu,* Strict Convexity and Complex Strict Convexity: Theory and Applications
90. *V. Lakshmikantham,* Trends in Theory and Practice of Nonlinear Differential Equations
91. *H. L. Manocha and J. B. Srivastava,* Algebra and Its Applications
92. *D. V. Chudnovsky and G. V. Chudnovsky,* Classical and Quantum Models and Arithmetic Problems
93. *J. W. Longley,* Least Squares Computations Using Orthogonalization Methods
94. *L. P. de Alcantara,* Mathematical Logic and Formal Systems
95. *C. E. Aull,* Rings of Continuous Functions
96. *R. Chuaqui,* Analysis, Geometry, and Probability
97. *L. Fuchs and L. Salce,* Modules Over Valuation Domains
98. *P. Fischer and W. R. Smith,* Chaos, Fractals, and Dynamics
99. *W. B. Powell and C. Tsinakis,* Ordered Algebraic Structures
100. *G. M. Rassias and T. M. Rassias,* Differential Geometry, Calculus of Variations, and Their Applications
101. *R.-E. Hoffmann and K. H. Hofmann,* Continuous Lattices and Their Applications
102. *J. H. Lightbourne, III, and S. M. Rankin, III,* Physical Mathematics and Nonlinear Partial Differential Equations
103. *C. A. Baker and L. M. Batten,* Finite Geometries
104. *J. W. Brewer, J. W. Bunce, and F. S. Van Vleck,* Linear Systems Over Commutative Rings
105. *C. McCrory and T. Shifrin,* Geometry and Topology: Manifolds, Varieties, and Knots
106. *D. W. Kueker, E. G. K. Lopez-Escobar, and C. H. Smith,* Mathematical Logic and Theoretical Computer Science
107. *B.-L. Lin and S. Simons,* Nonlinear and Convex Analysis: Proceedings in Honor of Ky Fan
108. *S. J. Lee,* Operator Methods for Optimal Control Problems
109. *V. Lakshmikantham,* Nonlinear Analysis and Applications
110. *S. F. McCormick,* Multigrid Methods: Theory, Applications, and Supercomputing
111. *M. C. Tangora,* Computers in Algebra
112. *D. V. Chudnovsky and G. V. Chudnovsky,* Search Theory: Some Recent Developments
113. *D. V. Chudnovsky and R. D. Jenks,* Computers and Mathematics
114. *M. C. Tangora,* Computers in Geometry and Topology

Other Volumes in Preparation

Computers in Geometry and Topology

edited by
Martin C. Tangora
University of Illinois at Chicago
Chicago, Illinois

Marcel Dekker, Inc. New York and Basel

Library of Congress Cataloging-in-Publication Data

Computers in geometry and topology / edited by Martin C. Tangora.
p. cm. -- (Pure and applied mathematics ; 114)
Proceedings of the Conference on Computers in Geometry and Topology, held Mar. 24–28, 1986, at the University of Illinois at Chicago.
Includes index.
ISBN 0-8247-8031-0
1. Geometry--Data processing--Congresses. 2. Topology--Data processing--Congresses. I. Tangora, Martin C. II. Conference on Computers in Geometry and Topology (1986 : University of Illinois at Chicago) III. Series: Monographs and textbooks in pure and applied mathematics ; v. 114.
QA448.D38C67 1989 88-26744
516'.028'5--dc19 CIP

This book is printed on acid-free paper.

MARCEL DEKKER, INC.
270 Madison Avenue, New York, New York 10016

Current printing (last digit):
10 9 8 7 6 5 4 3 2 1

PRINTED IN THE UNITED STATES OF AMERICA

Preface

Anyone who thinks that pure mathematics must be a sanctuary, forever insulated from the impact of modern computing machines, should read this volume.

These chapters show the extraordinary variety of ways in which the computer can help to solve old problems, to reformulate old problems or redirect energy and interest, and to pose new problems, in both old fields and new.

The range of subject matter is very broad, from the most visual geometry to the most algebraic topology. The text reflects finished research, work in progress, and exposition. The computers range from mainframes to laptops.

This book will be of interest, not only to the expert who wants to see the latest from Milnor on the Mandelbrot set, or from Curtis and Mahowald on the lambda algebra, but also to mathematicians who are just beginning to use a computer.

In the chapters by Davis, Harris, and Riley, the narrative tells about the exploratory process that each of these authors went through in trying to get various computers and various languages or software to help him find the right formulation of the theorem, the right reduction of the computation, or the simplest and most painless route to the results.

In the more geometrical chapters the power of computers to produce useful and attractive graphics is cogently demonstrated, and it is hard to imagine how the study of such constructs as the Mandelbrot set could continue without the constant support of the machine.

In the chapters that deal with algebraic topology and homological algebra the flavor is entirely different. Here, what David Eisenbud somewhat wistfully called the seductive world of intricate and beautiful graphics is replaced by the gritty detail of astronomically large computations in symbolic algebra.

At the computer-science end of the spectrum the reader will also find here two chapters on the design of mathematical software and one that applies complexity theory to algebraic topology.

All the papers arise from the Conference on Computers in Geometry and Topology, held on March 24—28, 1986, at the University of Illinois at Chicago. The conference was organized by the UIC Department of Mathematics, Statistics, and Computer Science, with Martin C. Tangora and Philip D. Wagreich as organizers. Sponsorship and financial support came from the UIC Graduate College, College of Liberal Arts and Sciences, and Office of Continuing Education, and from the National Science Foundation.

A majority of the talks of the invited speakers will be found here, some in revised form, along with several contributed papers from other participants. All the papers have been refereed.

Thanks are due to the UIC Computer Center, which, through its consistently liberal policy of free use for faculty and students, has fostered a campus-wide interest in exploring ways to bring the new technology to bear on problems of every kind.

The editor would like to express his gratitude to the referees; to Phil Wagreich, who provided indispensable support in the whole process of development of the book; to the electronic mail networks, BITNET and JANET, that saved numberless days of waiting for correspondence; and to the Mathematical Institute at Oxford—especially to Ioan James—for providing hospitality and facilities while the book was brought to completion.

Martin C. Tangora

Contributors

David J. Anick Associate Professor, Department of Mathematics, Massachusetts Institute of Technology, Cambridge, Massachusetts

Thomas F. Banchoff Professor, Department of Mathematics, Brown University, Providence, Rhode Island

Max Benson Associate Professor, Department of Computer Science, University of Minnesota, Duluth, Minnesota

Robert R. Bruner Assistant Professor, Department of Mathematics, Wayne State University, Detroit, Michigan

Edward Curtis Professor, Department of Mathematics, University of Washington, Seattle, Washington

Donald M. Davis Professor, Department of Mathematics, Lehigh University, Bethlehem, Pennsylvania

Peter J. Giblin Professor, Department of Mathematics, University of Massachusetts at Amherst, Amherst, Massachusetts, and Liverpool University, Liverpool, England

Ivan Handler Independent Software Developer, Chicago, Illinois

John C. Harris* Assistant Professor, Department of Mathematical Sciences, Purdue University Calumet, Hammond, Indiana

Louis H. Kauffman Professor, Department of Mathematics, Statistics, and Computer Science, University of Illinois at Chicago, Chicago, Illinois

*Current affiliation: Acting Assistant Professor, Department of Mathematics, University of Washington, Seattle, Washington

Larry A. Lambe* Professor, Department of Mathematics, North Carolina State University, Raleigh, North Carolina

Mark Mahowald Professor, Department of Mathematics, Northwestern University, Evanston, Illinois

John Milnor Professor, School of Mathematics, Institute for Advanced Study, Princeton, New Jersey

Douglas C. Ravenel Professor, Department of Mathematics, University of Rochester, Rochester, New York

David L. Rector Professor, Department of Mathematics, University of California at Irvine, Irvine, California

Robert F. Riley Assistant Professor, Department of Mathematical Sciences, State University of New York at Binghamton, Binghamton, New York

Dan Sandin Professor, Department of Art and Design, University of Illinois at Chicago, Chicago, Illinois

*Current affiliation: Professor of Mathematics and Computer Science, Department of Mathematics, Statistics, and Computer Science, University of Illinois at Chicago, Chicago, Illinois

Contents

Computers in Geometry and Topology

1

The Computation of Rational Homotopy Groups Is #𝒫-Hard

DAVID J. ANICK

Massachusetts Institute of Technology
Cambridge, Massachusetts

We give a measure of the computational complexity of homotopy groups. Given a finite simply connected CW complex X, a common problem in algebraic topology is to evaluate $\dim(\pi_n(X) \otimes Q)$. This problem is shown to belong to the class of #𝒫-hard problems, which are believed to require more than polynomial time to compute deterministically. Computing the Hilbert series of a graded algebra or the Poincaré series of a local Artinian ring is also #𝒫-hard.

Intended for topologists, the exposition is self-contained, assuming no prior familiarity with theoretical computer science concepts.

1. HISTORICAL CONTEXT

Beginning with the earliest definitions of homotopy groups $\pi_*(\cdot)$ by Čech [10] and Hurewicz [16], topologists have been interested in computing or describing them. In [25] Serre recognized that the homotopy groups of a finite 1-connected simplicial complex X would be finitely generated abelian groups. By the known classification of such groups, any $\pi_n(X)$ would have to have a simple description as

$$\pi_n(X) = \underbrace{Z \oplus \cdots \oplus Z}_{r} \oplus Z_{a_1} \oplus \cdots \oplus Z_{a_m} \qquad [1]$$

where $a_1 \mid a_2 \mid \cdots \mid a_m$. This discovery meant that $\pi_*(\cdot)$ could be be viewed as a function, with a finite description of X and an integer n as inputs and with the finite list $(r; a_1, a_2, \ldots, a_m)$ as output.

In 1957 Brown [9] proved that there exists an algorithmic procedure, which always terminates after a finite number of steps, for evaluating this function. At the time this seemed to settle the matter: homotopy groups can be computed, and topologists' job is to do it. However, despite much creative work since then, our explicit knowledge of homotopy groups remains dismally poor. As Paul Selick has observed, there is not a single noncontractible finite 1-connected CW complex X for which $\pi_n(X)$ is known for all n. Experience teaches us that to call the computation of general homotopy groups "very hard" would be a laughable understatement.

A crucial ray of light in this darkness was offered by Quillen in 1969 and later expanded by Sullivan [27] and others. Quillen [23] showed that the integers r of (1), corresponding to the ranks of the rational homotopy groups $\pi_*(X) \otimes Q$, participate in an algebraic model for X_Q (X_Q is X localized at Q) which renders them far more accessible than the whole $\pi_*(X)$. By using this model, the complete rational homotopy of X is now known for a great many finite 1-connected X. Rational homotopy seems to be significantly easier to compute than the total homotopy group, and developments in rational homotopy continue to shed light on questions about $\pi_*(X)$.

During the past 15 years the field of theoretical computer science has blossomed tremendously, from a near obscurity practiced by a few specialists to a major mathematical effort involving many deep concepts and theorems. Questions about the computability of algebraic or topological objects can now be posed in far more refined ways, and the answers will have both theoretical and practical importance.

This, then, is the context in which the results of this chapter appear. Computation of the free abelian component of $\pi_n(X)$ will be shown to be $\#\mathcal{P}$-hard, a well-known concept in computer science. Since $\#\mathcal{P}$-hard problems sit rather high on the complexity scale, this confirms our experience that homotopy groups, even rational homotopy groups, are difficult to evaluate.

Computer scientists conjecture, but have not proved, that $\#\mathcal{P}$-hard problems cannot be solved in polynomial time on an ordinary deterministic machine. If true, this conjecture would imply that a worst-case rational homotopy calculation requires more than polynomial time, which would be a very powerful result. Thus topologists are left in the position of waiting until further advances are made in the field of theoretical computer science!

Even if this conjecture were solved, it would not close the chapter on computability in homotopy. For one thing, there has been increased interest recently in extending Brown's results to non-simply connected spaces. Weld [29] succeeded at this for nilpotent CW complexes. Her results suggest that the best we can hope for with a typical nilpotent space X will be a recursively enumerable presentation for $\pi_n(X)$. Whether one can obtain $\pi_n(X_Q)$ precisely for nilpotent X and whether Brown's results can be extended to the determination of [Y; X] when Y is not a sphere and $\overline{H}_*(X)$ is not a torsion group remain for future research to reveal.

2. SUMMARY OF RESULTS

In computer science, a "problem" is a function f from a subset of N, the nonnegative integers, to N. Sometimes the argument of this function, also called the "input," may be viewed naturally as a single integer, while at other times it may encode, through some fixed injection $\eta : \amalg_{j=1}^{\infty} Z^j \to N$, the finite description of some other object. (The map η is said to N-encode the finite description.) Regardless of the proper interpretation of the input, a computer scientist may imagine that a machine or algorithm exists which can accept an arbitrary $N \in Dom(f)$ as input and deliver f(N) as output, utilizing in the process some number T(N) of steps. Given f, he or she may seek theoretical lower bounds for the function T(N) or seek algorithms (machines) on which T(N) exhibits a certain level of efficiency.

Computer scientists have developed a scale or continuum along which various problems may be placed according to their complexity. Certain classes of problems, including the classes known as $\mathcal{P}$, #$\mathcal{P}$-complete, and "computable in exponential time," serve as landmarks. In particular, there is a transitive, reflexive relation on the set of problems, called "Turing reducible in polynomial time" and denoted $\underset{T}{\leq}$, by which f_2 is at least as hard as f_1 if $f_1 \underset{T}{\leq} f_2$. The problems f_1 and f_2 are "Turing equivalent," denoted $f_1 \underset{T}{\approx} f_2$, if if $f_1 \underset{T}{\leq} f_2$ and $f_2 \underset{T}{\leq} f_1$.

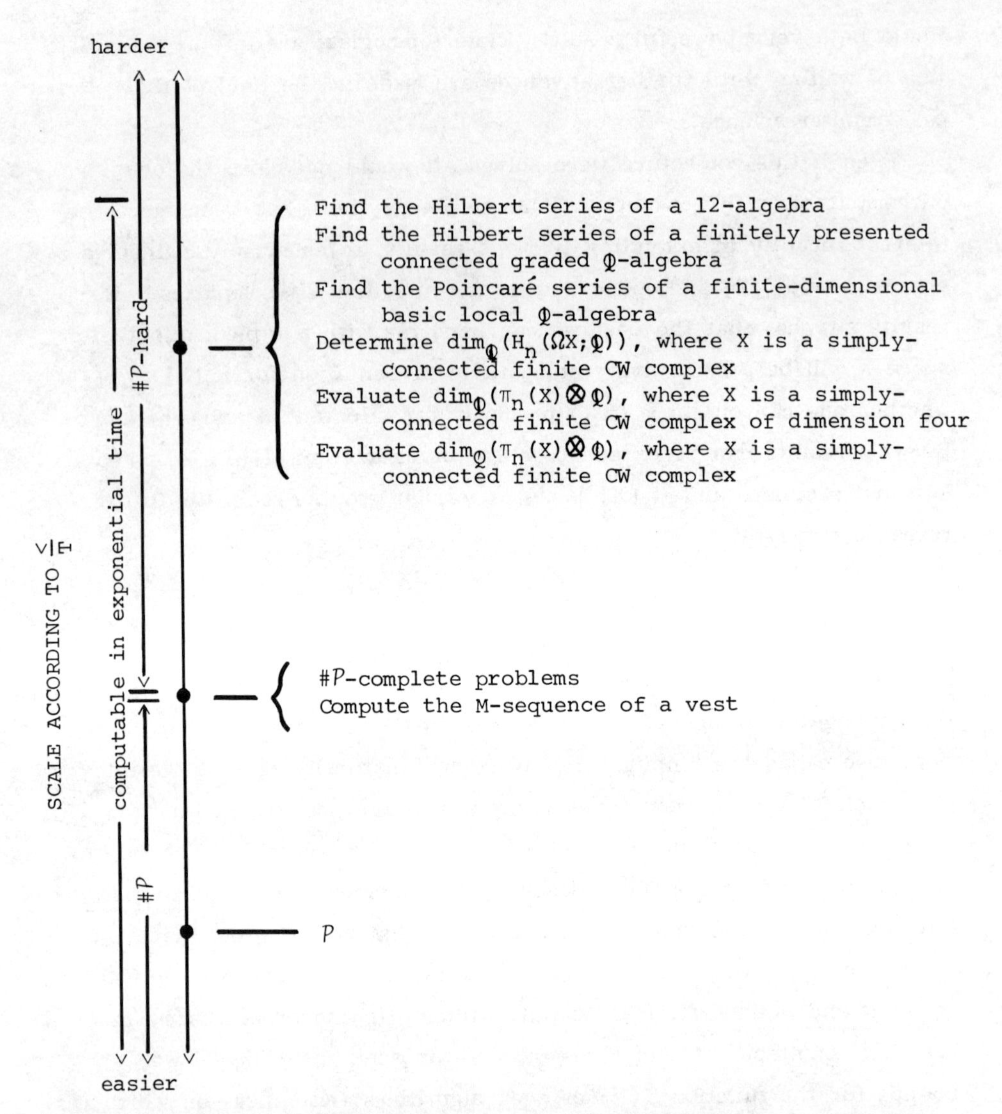

Figure 1 Diagram showing a portion of the computational complexity scale and the relative positions of three Turing equivalence classes.

The goal of this chapter is to locate the problem "compute rational homotopy" and some related problems on this continuum.

Our results are summarized in Figure 1. More difficult problems are placed higher on the scale, and Turing equivalent problems are bracketed. Problems at a specific difficulty level are marked by filled circles on the vertical axis. Classes of problems which encompass a range along the continuum are exhibited as intervals.

The purpose of Figure 1 is to give a visual overview of our results, and some technical points were sacrificed for crispness. The figure may be misleading in that the partial order $\underset{T}{\leq}$ is depicted as a linear order. For instance, one might conclude from the figure that any problem unsolvable in exponential time is #$\mathcal{P}$-hard, but this has not been proved (and is probably false). Nor have we proved that the three Turing equivalence classes marked by filled circles must be actually distinct.

As to interpretation, bear in mind that the class #$\mathcal{P}$-hard starts near the bottom of Figure 1, even though this class is viewed by computer scientists as being "very difficult." In other words, except for "$\mathcal{P}$," the section of the scale shown in Figure 1 actually starts far above such familiar computer mainstays as "solve a linear system of equations," "find the roots of a polynomial," or "factor the integer N."

There are no known algorithms which can evaluate a #$\mathcal{P}$-hard problem in less than exponential time. Algorithms which require exponential time are generally thought of as being beyond the scope of today's machinery to implement efficiently. Thus computing rational homotopy groups and the other problems listed in Figure 1 may truly be described as very complex problems.

3. THE CONNECTION WITH HILBERT SERIES

The computational complexity of rational homotopy groups can best be discerned by studying certain closely related calculations which involve graded algebras. In this section we shall explore this connection. Finding the rational homotopy groups of a space is computationally equivalent to evaluating something we will call "Tor-sequences." These in turn contain as a subset the collection of "M-sequences." We close with some interesting examples of M-sequences in order to illustrate how complicated rational homotopy can be.

For the remainder of this chapter, a space will refer to a finite simply connected CW complex whose 1-skeleton is trivial.

In studying the computational complexity of homotopy groups, a certain technical sticking point arises immediately. As we have noted, $\pi_*(\cdot)$ may be viewed as a function whose inputs or arguments are an integer n and a description of a space X. How, exactly, does one describe a space?

One possible answer was adopted by Brown. He assumed that the description of X would consist of a simplicial decomposition, i.e., a list of the simplices for a simplicial complex X' having the same homotopy type as X. Such a list, however, is extremely long for any space of even moderate complexity. For example, several hundred simplices are involved in the smallest such description of $S^2 \times S^2$. Ideally, one would like to describe a CW complex in such a way that the length of the description is roughly comparable to the number of cells and/or the complexity of the various attaching maps.

Fortunately, in the case of a rational homotopy, Quillen's Lie algebra model $(\mathcal{L}X, d_X)$ for a space X does provide such a description. Any space $X = * \cup (\cup_{i=1}^{h} e^{m_i})$ (assume $m_i \geq 2$) can be specified up to rational homotopy type by giving a nondecreasing list $(m_1, \ldots, m_h)$ of the degrees in which the cells occur followed by a description of the differential d_X. Since $\mathcal{L}X$ can be taken to be the free graded Lie Q-algebra on generators $\{x_1, \ldots, x_h\}$, where $\deg(x_i) = m_i - 1$, describing d_X amounts to giving m homogeneous elements of $\mathcal{L}X$. The ith entry in this list should have degree $m_i - 2$. Thus for rational homotopy calculations the <u>description</u> of a space will consist of a list $(m_1, \ldots, m_h)$ of cell dimensions, followed by a list $(d_X(x_1), \ldots, d_X(x_h))$ of boundaries. For example, when $S^2 \times S^2$ has its usual CW decomposition its description could be

$$(2, 2, 4); (0, 0, [x_1, x_2]) \qquad [2]$$

In Section 5 we specify a precise form for this description.

Three interrelated points need to be made. First, a valid description of a space is always obtainable from a simplicial decomposition, so we are justified in assuming that the description is available as input. Second, we shall want the input to be comparable in length to the "size" or "complexity" of X, and for this the above notion of description works well. Third, one can easily N-encode a string of symbols such as (2) into a natural number whose length (when written, say, in decimal) is bounded by a constant multiple of the total number of symbols in the original description.

We will henceforth view the problem of computing rational homotopy as the problem of using a valid Quillen model (equivalently, a valid description or N-encoded description) and an integer n to generate the integer.

$$r_n(X) = \dim(\pi_{n+1}(X) \otimes Q) \qquad [3]$$

The theorem which makes this feasible and which motivates the dimension shift in (3) is

Theorem 3.1 (Quillen)

If $(\mathcal{L}X, d_X)$ is the Quillen model for a space X, then $d_X^2 = 0$ and

$$H_q(\mathcal{L}X, d_X) \approx \pi_q(\Omega X) \otimes Q$$

Equivalently,

$$\pi_{q+1}(X) \otimes Q = \ker(d_X)_{(q)}/\mathrm{im}(d_X)_{(q+1)}$$

enclosed subscripts denoting graded components, or

$$r_n(X) = \dim(\ker(d_X)_{(n)}) - \dim(\mathrm{im}(d_X)_{(n+1)}) \qquad [4]$$

We shall see that it is easier to work with the Betti sequence of ΩX than with $\{r_n(X)\}$. The Betti sequence of ΩX is the sequence $\{b_n(\Omega X)\}$, where $b_n(\Omega X) = \dim(H_n(\Omega X; Q))$. By [22] these sequences are related by the formula

$$\sum_{i=0}^{\infty} b_i z^i = \prod_{j=1}^{\infty} \frac{(1 + z^{2j-1})^{r_{2j-1}}}{(1 - z^{2j})^{r_{2j}}} \qquad [5]$$

where $b_i = b_i(\Omega X)$ and $r_j = r_j(X)$. The left-hand side of (5) is called the Poincaré series of ΩX and will be denoted by $P_{\Omega X}(z)$.

By formula (5), knowing either one of $\{b_i\}_{i \leq n}$ or $\{r_j\}_{j \leq n}$ enables us to compute the other. Furthermore, the computation involved is a one which computer scientists would think of as being executable "quickly." "Fast" calculations are those which require only a polynomial number of steps. Let us see why this computation qualifies.

To obtain $\{b_i\}_{i\leq n}$ when one has the list $\{r_i\}_{i\leq n}$, one can view (5) as a congruence modulo z^{n+1}. To evaluate the right-hand side of (5) requires that we multiply together the polynomials

$$(1+z)^{r_1},\ (1+z^2+z^4+\cdots+z^{\bar{n}})^{r_2},\ (1+z^3)^{r_3},\ (1+z^4+z^8+\cdots+z^{\bar{n}})^{r_4},\ldots \quad [6]$$

where "$z^{\bar{n}}$" is used loosely to denote "stop after z^n." To multiply together two polynomials of degree n (modulo z^{n+1}) takes at most $1 + 2 + \cdots + (n + 1)$ multiplications and $0 + 1 + \cdots + (n)$ additions, which we summarize as $(n + 1)^2$ operations. To raise anything to the power r requires at most $2\cdot\log_2(r)$ multiplications; to see this, write r in binary and use the trick that $x \to x^2 \to x^4 \to \cdots \to x^{2^m}$ takes only m multiplications. Thus at most

$$(n+1)^2\,(2)\,[\log_2(r_1) + \log_2(r_2) + \cdots + \log_2(r_n)]$$

operations are needed in order to evaluate the entries of the list (6), and at most $(n + 1)^2(n - 1)$ further operations are involved in forming their product (modulo z^{n+1}).

Finally, one sees easily for each space X that $\{r_j(X)\}$ grows at most exponentially with j, that is, $\log_2(r_j) \leq Kj$ for some fixed K. [An upper bound for K is $\log_2(m)$ if X has m cells.] Evaluating the entire right-hand side of (5) takes at most

$$(n+1)^2(n-1) + (n+1)^2\,(2)\,K\left(\frac{n^2+n}{2}\right)$$

operations. Since this is a polynomial in n, it justifies the description of this calculation as "fast."

Likewise, one can "quickly" obtain $\{r_j\}_{j\leq n}$ from knowledge of $\{b_i\}_{i\leq n}$. The problem of computing $\{r_j\}_{j\leq n}$ and the problem of computing $\{b_i\}_{i\leq n}$ are viewed as being "equivalent" to one another. We postpone the precise definition of this equivalence until Section 5. For the time being, we hope the reader is convinced that it suffices to study the complexity of $\{b_n(\Omega X)\}$ in order to understand the problem of computing $\{r_n(X)\}$.

Remark. For the purposes of implementation, there are more efficient ways to multiply two polynomials than the naive way analyzed above. See [24, chapters 4 and 36] for an excellent discussion of these faster algorithms.

To determine $\{b_i(\Omega X)\}_{i \leq n}$ is the same as to determine the Poincaré series of ΩX modulo z^{n+1}. This in turn is equivalent to evaluating the first n coefficients of the Hilbert series of a certain graded algebra. We therefore detour to introduce some terminology related to graded algebras.

In this chapter, a graded algebra is a connected N-graded finitely presented Q-algebra A. Such an object always has a presentation as

$$A \approx Q\langle x_1,\ldots,x_g\rangle/\langle\alpha_1,\ldots,\alpha_r\rangle \qquad [7]$$

In (7), $Q\langle x_1,\ldots,x_g\rangle$ is the free associative Q-algebra on $\{x_1,\ldots,x_g\}$, which is N-graded by assigning a positive integral degree $|x_i|$ to each x_i. The notation $\langle\alpha_1,\ldots,\alpha_r\rangle$ designates the two-sided ideal of $Q\langle x_1,\ldots,x_g\rangle$ generated by the finite set of homogeneous relations $\{\alpha_1,\ldots,\alpha_r\}$. Since $\langle\alpha_1,\ldots,\alpha_r\rangle$ is homogeneously generated, the quotient algebra A inherits a gradation from $Q\langle x_1,\ldots,x_g\rangle$ and we may write $A = \oplus_{j=0}^{\infty} A_j$. The Hilbert sequence of A is the sequence $\{h_j(A)\}$, where $h_j(A) = \dim_Q(A_j)$, and the Hilbert series is $H_A(z) = \sum_{j=0}^{\infty} h_j(A)z^j$.

A one-two algebra (or 12-algebra) is a graded algebra A which has a presentation (7) such that each $|x_i| = 1$ and each $|\alpha_k| = 2$. A 123-algebra is a 12-algebra which has global dimension three. A 12H-algebra is a 12-algebra which is also a Hopf algebra. This is equivalent to the condition that each α_k be a Q-linear combination of $(x_ix_j + x_jx_i)$'s and (x_i^2)'s. These were called Roos algebras in [3]. Lastly, a 123H-algebra is a 12H-algebra of global dimension three.

The connection between graded algebras and arbitrary spaces is made in [6]. By Theorem 1 of [6] there exists for every space X a 123H-algebra A such that $H_A(z)$ and $P_{\Omega X}(z)$ are rationally related. This means that there exist polynomials $p_i(z) \in Z[z]$, $1 \leq i \leq 4$, such that $p_1p_4 \neq p_2p_3$ and

$$P_{\Omega X}(z) = \frac{p_1(z)H_A(z) + p_2(z)}{p_3(z)H_A(z) + p_4(z)} \qquad [8]$$

Conversely, for any 12-algebra A, there exists a space X of dimension four for which (8) holds.

Formula (8) implies that the problem of computing $\{b_i(\Omega X)\}_{i\leq n}$ and the problem of determining $\{h_j(A)\}_{j\leq n}$ are equivalent, in the sense that one could obtain either list from the other after only q(n) additional operations for some fixed polynomial q. To see this, note that in passing from H_A to $P_{\Omega X}$ the only possible trouble spot is the inversion of $P_3(z)H_A(z) + p_4(z)$. By the proof of Theorem 1 of [6], however, we can assume that $p_3(z)H_A(z) + p_4(z) \equiv 1 (\mathrm{mod}\ z)$ and we may write

$$p_3(z)H_A(z) + p_4(z) = 1 - z\cdot u(z), \qquad u(z) \in Z[z]$$

Then $(1 - zu(z))^{-1} = 1 + zu(z) + z^2u(z)^2 + \cdots$ and

$$(1 - zu(z))^{-1} \equiv \theta^n(1)\ (\mathrm{mod}\ z^{n+1})$$

where θ is the endomorphism of Z[z] defined by $\theta(x) = 1 + zu(z)x$. Since evaluating $\theta(x)$ (mod z^{n+1}) takes up to $(n + 1)^2$ operations, we have inverted $p_3(z)H_A(z) + p_4(z)$ in only $n(n + 1)^2$ operations.

Thus the general problem of computing $\{r_j(X)\}_{j\leq n}$ is equivalent to the problem of finding $\{h_j(A)\}_{j\leq n}$ for a certain 123H-algebra A associated to X.

Remark. In practice, there are more efficient ways to take the quotient of two power series. See, for example, Sections 4.6.1 and 4.7 of [18].

We perform just one more reduction. Because gl.dim(A) = 3, its Hilbert series can be expressed neatly in terms of another series, which we will call its "Tor-series." Specifically, note that $\mathrm{Tor}_3^A(Q, Q)$ is a graded Q-module because A is graded, and let $c_q = c_q(A) = \dim(\mathrm{Tor}^A_{3,q+3}(Q, Q))$. The Tor-sequence of A is $\{c_q\}_{q\geq 0}$ and the Tor-series of A is $T_A(z) = \sum_{q=0}^{\infty} c_q z^q$. Because A has global dimension three,

$$H_A(z)^{-1} = 1 - gz + rz^2 - z^3T_A(z) \qquad [9]$$

Here g and r count the numbers of generators and relations in a minimal presentation (7) for A.

On the basis of the previous remarks about multiplying and inverting power series, we assert without further proof that computing $\{r_i(X)\}_{i \leq n}$ is equivalent to computing the Tor-sequence $\{c_q\}_{q \leq n-3}$ of a certain 123H-algebra A associated to X.

We have gotten rather far afield from rational homotopy groups, but Tor-sequences are a good place to stop. Tor-sequences capture what is intrinsically complicated about rational homotopy but express that intricacy in more accessible algebraic or combinatorial terms. A further advantage of Tor-sequences is that we may more readily construct bizarre or amusing examples to illustrate their diversity. Lastly, the class contains a special subset, to be called "M-sequences," which is natural in the following sense. The general problem of "compute an M-sequence" is #$\mathcal{P}$-complete, an important computer science concept.

To summarize what we have shown so far, we have

Proposition 3.2

For every space X there exists an associated 123H-algebra A with the following property. Given the first n terms of any one of the following sequences, one can with $\tau(n)$ additional operations compute the first n terms of any other sequence, where $\tau(n)$ is a polynomial in n:

1. The rational homotopy ranks $r_j(X) = \dim(\pi_{j+1}(X) \otimes Q)$
2. The Betti sequence $b_i(\Omega X) = \dim(H_i(\Omega X; Q))$
3. The Hilbert sequence $h_j(A) = \dim(A_j)$
4. The Tor-sequence $c_i(A) = \dim(\mathrm{Tor}^A_{3,3+i}(Q, Q))$

Conversely, given any 123-algebra A, there exists a space X with $\dim(X) = 4$ for which the same conclusion holds.

We will now jump right in with the definition of an M-sequence. For motivation, the reader is welcome to glance ahead to Theorem 3.4.

Definition 3.3

A vector evaluated after a sequence of transformations, henceforth vest, is a four-tuple (d, v_0, T, S) as follows. The first entry d is any positive integer, and v_0 is any vector, called the initial vector, in Q^d. The third entry T denotes a list $T_1, \ldots, T_m$ of $d \times d$ matrices over Q viewed as linear transformations on Q^d, and S is any $h \times d$ matrix over Q. Given a vest (d, v_0, T, S) and a length n sequence of indices $\sigma = (i_1, \ldots, i_n)$ having $1 \leq i_j \leq m = \#(T)$ define $v_0 * \sigma$ to be the vector

$$v_0 * \sigma = T_{i_n} \cdots T_{i_2} T_{i_1}(v_0) \in Q^d$$

Let $M_n = \{\sigma = (i_1, \ldots, i_n) \mid S(v_0 * \sigma) = 0\}$ and let $e_n = \#(M_n)$. The <u>M-sequence</u> for (d, v_0, T, S) is $\{e_n\}_{n \geq 0}$ and its <u>M-series</u> is the formal power series $M(z) = \sum_{n=0}^{\infty} e_n z^n$. An arbitrary formal power series M(z) is an <u>M-series</u> if and only if it equals the M-series of some vest.

Note that a vest can be specified by listing an integer d, d rational numbers, an integer m, md^2 more rational numbers, and the integer h followed by hd rationals. Note that the M-sequence is not affected if v_0 or S or any T_i is multiplied by a nonzero scalar, so no harm is done by clearing denominators and assuming that all entries are integers. A vest can therefore be thought of as being specified by a list of integers, the length of the list being $1 + d + 1 + md^2 + 1 + hd$.

The motivation for considering Definition 3.3 lies in the following theorem, which translates Theorem 1.3 of [3] into the language of that definition.

Theorem 3.4

Let (d, v_0, T, S) be a vest and let M(z) be its M-series. Then there exists a 123H-algebra A, with $g = 2m + d + h + 3$ generators and $r = (m + 1)(m + d + h + 2) + 1$ relations, whose Tor-sequence equals M(z). In other words, every M-sequence is a Tor-sequence.

Since every M-sequence is a Tor-sequence, Tor-sequences must be at least as difficult to compute, in general, as M-sequences. Since Tor-sequences are comparable in computational complexity to rational homotopy groups, rational homotopy is at least as hard to calculate as M-sequences. We shall see in the next section that general M-sequences are as hard as or harder to compute than any problem belonging to a large class called #$\mathcal{P}$. The remainder of this section is devoted to some examples of M-sequences which we hope will illustrate how wide a class of functions they encompass. It can be skipped by the reader without loss of continuity.

Theorem 3.5

Fix positive integers m and h. Let $\mathfrak{a}$ denote the m-tuple $(a_1,\ldots,a_m) \in \mathrm{N}^m$. For $1 \leq i \leq h$ let $E_i(n,\mathfrak{a})$ be a Q-linear combination of expressions of the form

$$c^n n^{d_0} a_1^{d_1} \cdots a_m^{d_m} \qquad [10]$$

where $c \in Q - \{0\}$ and $d_j \in \mathrm{N}$. Let $B_1,\ldots,B_m \in \mathrm{N} - \{0,1\}$ be arbitrary and put

$$I_n = \bigcap_{i=1}^{h} \{\mathfrak{a} = (a_1,\ldots,a_m) \in \mathrm{N}^m \mid E_i(n,\mathfrak{a}) = 0 \text{ and } 0 \leq a_j < B_j^n \text{ for } 1 \leq j \leq m\}$$

Then the list of cardinalities $\{\#(I_n)\}_{n \geq 0}$ is an M-sequence.

The proof is postponed until after two motivating examples are given.

Example 3.6. The sequence $e_n = [(3/2)^n]$, brackets denoting the greatest integer function, is an M-sequence.

Proof. Consider the single equation

$$E(a_1, a_2) = 2^n(a_1 + 1) + a_2 - 3^n = 0 \qquad [11]$$

and put $B_1 = 2$, $B_2 = 3$. Nonnegative integral solutions to (11) which satisfy the bounds $0 \leq a_1 < 2^n$ and $0 \leq a_2 < 3^n$ are in one-to-one correspondence with

$$\{a_1 \in \mathrm{N} \mid 2^n(a_1 + 1) \leq 3^n\}$$

This set has cardinality e_n.

Example 3.7. The sequence $\{e_n\}$, where e_n equals the nth digit in the decimal representation of $\sqrt{2}$, is an M-sequence.

Proof. Consider the system of h = 4 equations:

$$E_1(n,\mathcal{a}) = (10a_1 + a_2)^2 + a_3 - 2(100)^n = 0$$

$$E_2(n,\mathcal{a}) = (10a_1 + a_2 + 1)^2 - a_4 - 2(100)^n = 0$$

$$E_3(n,\mathcal{a}) = a_2 + a_5 - 10 + 1 = 0$$

$$E_4(n,\mathcal{a}) = a_6 + a_7 - a_2 + 1 = 0$$

and seek simultaneous solutions subject to the bounds $B_3 = B_4 = 100$, $B_1 = B_2 = B_5 = B_6 = B_7 = 10$.

Putting $x = 10a_1 + a_2$, the first two equations say that

$$x^2 \leq 2(10)^{2n} \leq (x+1)^2$$

so any solution has $x = [10^n\sqrt{2}]$. The third equation assures us that $0 \leq a_2 < 10$, which forces a_2 to equal the last decimal digit of x; clearly this digit is e_n. So far, there is a unique solution for (a_1, a_2, a_3, a_4, a_5). The fourth equation permits the total number of simultaneous solutions for $\mathcal{a}$ to equal $a_2 = e_n$. Thus $\#(I_n) = e_n$, as desired.

Proof of Theorem 3.5. Given a system of expressions $\{E_j(n,\mathcal{a})\}$ in which each $E_i(n,\mathcal{a})$ is assumed to be a linear combination of terms (10), we want to construct a vest (δ, v_0, T, S) whose M-sequence measures the number of suitably bounded solutions to "$E(n,\mathcal{a}) = 0$." We will do this by letting T consist of $B_1B_2\cdots B_m$ linear transformations denoted T_λ, where λ runs through the set of m-tuples

$$\Lambda = \{\lambda = (\lambda_1,\ldots,\lambda_m) \in \mathbb{N}^m \mid 0 \leq \lambda_j < B_j\}$$

We set up a bijection $\mathcal{G}$ between the set Λ^n of length n sequences $\sigma = (\lambda(1),\ldots,\lambda(n))$ and the set of m-tuples

$$\mathcal{A}(n) = \{\mathcal{a} = (a_1,\ldots,a_m) \in \mathbb{N}^m \mid 0 \leq a_j < B_j^n\}$$

in the following manner. The $(n-i)$th digit, in base B_j, of a_j will be the jth entry of the m-tuple $\lambda(i)$. The m-tuple $\mathcal{a}$ obtained in this way from a sequence $\sigma \in \Lambda^n$ will be denoted $\mathcal{G}(\sigma)$.

Let $L_1(n,\mathcal{a}),\ldots,L_t(n,\mathcal{a})$ be a complete list of all the terms (10) which appear in any of the $E_i(n,\mathcal{a})$'s. For $1 \leq j \leq t$ we claim the existence of a vector space V_j and a $v_j \in V_j$, together with an action of each T_λ on V_j, such that for any $\sigma = (\lambda(1),\ldots,\lambda(n)) \in \Lambda^n$ the first component of the vector $v_j * \sigma$ equals $L_j(n, \mathcal{G}(\sigma))$.

Granting this claim, put $V = \oplus_{j=1}^{t} V_j$ and $v_0 = (v_1,\ldots,v_t) \in V$. Then each $E_i(n, \mathcal{G}(\sigma))$ equals a linear combination of certain components of $v_0 * \sigma$. The matrix S may be chosen so that the inner product of the ith row of S with $v_0 * \sigma$ equals $E_i(n, \mathcal{G}(\sigma))$. Then

$$S(v_0 * \sigma) = E(n, \mathcal{G}(\sigma))$$

in Q^h. As σ runs through Λ^n, $\mathcal{G}(\sigma)$ runs through $\mathscr{A}(n)$ exactly once. Thus $\#(I_n) = \#\{\sigma \mid S(v_0 * \sigma) = 0\}$, proving $\{\#(I_n)\}$ to be the M-sequence for the vest $(\dim(V), v_0, T, S)$.

We will prove the claim by induction on $d = d_0 + d_1 + \cdots + d_n$, where in keeping with (10) we write $L_j(n, \mathfrak{a}) = c^n n^{d_0} a_1^{d_1} \cdots a_m^{d_m}$. If $d = 0$ then $d_0 = d_1 = \cdots = d_m = 0$ and $L_j(n, \mathfrak{a}) = c^n$, so we may take $V_j = Q$ with $v_j = 1$ and take every T_λ to be multiplication by c.

Assuming the claim has been proved for exponent sums less than d, suppose that $d_0 + \cdots + d_m = d > 0$. For $\lambda = (\lambda_1,\ldots,\lambda_m) \in \Lambda$ and $\mathfrak{a} = (a_1,\ldots,a_m) \in \mathscr{A}(n)$, let $\mathfrak{a} * \lambda$ denote

$$\mathfrak{a} * \lambda = (B_1 a_1 + \lambda_1,\ B_2 a_2 + \lambda_2, \ldots, B_m a_m + \lambda_m)$$

The binomial formula applied to $L_j(n + 1, \mathfrak{a} * \lambda)$ shows

$$L_j(n + 1, \mathfrak{a} * \lambda) = (c B_1^{d_1} \cdots B_m^{d_m}) L_j(n, \mathfrak{a}) + E_j'(n, \mathfrak{a})$$

where $E_j'(n, \mathfrak{a})$ is a linear combination of expressions of the form (10) whose exponent sums are smaller than d. By our inductive hypothesis there exists a vector space V_j' on which transformations T_λ' act and an initial vector v_j' for which

$$S_j'(v_j' * \sigma) = E_j'(n, \mathcal{G}(\sigma))$$

for some row vector S_j'. Let $V_j = Q \oplus V_j'$ with

$$v_j = \begin{bmatrix} 0 \\ v_j' \end{bmatrix}$$

and let T_λ be the matrix

$$\begin{bmatrix} c_j & S'_j \\ 0 & T'_\lambda \end{bmatrix}$$

where $c_j = cB_1^{d_1} \cdots B_m^{d_m}$. An induction on n shows that the first component of $v_j * \sigma$ is indeed $L_j(n, \mathcal{G}(\sigma))$ for $\sigma \in \Lambda^n$. This completes the proof.

Remark. For further interesting examples of M-sequences the reader is referred to Corollary 2.5 of [3].

4. COMPUTING M-SEQUENCES IS # $\mathcal{P}$-COMPLETE

Section 4 is the technical core of the chapter, and it contains two major results. We first review the definitions of the terms $\mathcal{P}$, #$\mathcal{P}$, and #$\mathcal{P}$-complete. These are classes of problems which are solvable by various sorts of Turing machines. The first major result, Theorem 4.5, asserts that computing the M-sequence of a vest (see Definition 3.3) is #$\mathcal{P}$-complete. We indicate precisely how a vest should be N-encoded for this result to be true. The second major theorem, Theorem 4.8, proves a wide class of functions to be M-sequences. Membership in this class is contingent on computability by a $K \cdot (n)^{\varepsilon}$-time algorithm, a criterion satisfied by many familiar functions.

We begin by reviewing the concept of a Turing machine, for which there are several equivalent formulations. We adopt a variation on the description found in [26, p. vii], which is recalled next.

> Let there be given a tape of infinite length [in both directions] which is divided into squares and a finite list of symbols which may be written on these squares. There is an additional mechanism, the head, which may read the symbol on a square, replace it by another or the same symbol and move to the adjoining square to the left or right. This is accomplished as follows: At any given time the head is in one of a finite number of internal states. When it reads a square it prints a new symbol, goes into a new internal state and moves to the right or left depending on the original internal state and the symbol read. Thus a Turing machine is described by a finite list of quintuplets such as 3,4,3,6,R which means: If the machine is in the third internal state and reads the fourth symbol

it prints the third symbol, goes into the sixth internal state and moves to the right on the tape.

Let U denote the (finite) set of internal states. We specify an initial state $u^* \in U$ and two possible "terminal states" u_0 and u_1. Let $\hat{U} = U - \{u_0, u_1\}$ consist of the nonterminal states.

The initial configuration of the tape is assumed to consist of a single contiguous finite segment of nonblank squares with the head initially positioned immediately to the right of the rightmost nonblank symbol. This contiguous segment is the <u>input</u> I, whose <u>length</u>, denoted $\ell(I)$, is the number of nonblank squares. Without loss of generality we identify the set of tape symbols with $[B] = \{0,1,2,\ldots,B - 1\}$ for some integer $B \geq 4$, with "zero" being viewed as the "blank." The next B_0 symbols $(2 \leq B_0 \leq B - 2)$ are viewed as the digits 0 through $B_0 - 1$ in some base B_0, and among the remaining symbols we reserve one, called "semicolon," to delimit various segments of the input.

As noted above, the Turing machine itself consists of a collection Γ of quintuplets,

$$\Gamma \subseteq \hat{U} \times [B] \times [B] \times U \times \{+1,-1\}$$

where we are assuming that computation ceases if ever the machine attains a terminal state. Thus Γ may be viewed as a relation between the sets $\hat{U} \times [B]$ and $[B] \times U \times \{\pm 1\}$. In an ordinary <u>deterministic</u> Turing machine, the relation Γ is a function, and each configuration in $\hat{U} \times [B]$ leads to exactly one successive configuration (as indicated by an element of $[B] \times U \times \{\pm 1\}$). Deterministic machines embody the usual concept of predictable, reproducible, serial calculation. For a given input there is just one possible flow path through the various configurations.

On the other hand, a <u>nondeterministic</u> Turing machine is one for which the set Γ need not be a function. We do assume, for each $(u,b) \in \hat{U} \times [B]$, that there is at least one $(b',u',z') \in [B] \times U \times \{\pm 1\}$ such that $(u,b,b',u',z') \in \Gamma$. For each input I there could be many possible flow paths which a nondeterministic machine could follow, and some paths may end in each of the two terminal states. Nondeterministic machines are different from but closely related to probabilistic machines, which are computers that can incorporate the output of a random number generator into their branching decisions.

Definition 4.1

A function $f : \mathrm{Dom}(f) \to \mathbf{N}$, $\mathrm{Dom}(f) \subseteq \mathbf{N}$, is in the class $\mathcal{P}$ if and only if there exists a deterministic Turing machine, together with a polynomial $\tau(x)$, with the following property. Whenever the input is $N \in \mathrm{Dom}(f)$, the Turing machine will attain the terminal state u_0 after at most $\tau(\ell(N))$ steps, and before attaining the state u_0 the output $f(N)$ will be printed on the tape. Without loss of generality we assume that the machine terminates only after repositioning the head to the right of the rightmost nonblank square.

Example. The problem of multiplying two numbers is in $\mathcal{P}$. Given N_1 and N_2, we may $\mathbf{N}$-encode the pair (N_1,N_2) as

$$I = (N_1)_{(10)};(N_2)_{(10)}$$

where $(N)_{(10)}$ denotes the base 10 representation of the number N. (For convenience we have taken $B_0 = 10$.) By way of illustration, notice that to multiply $N_1 = 11374$ by $N_2 = 1286$ takes $5 \cdot 4 = 20$ individual multiplication operations followed by a comparable number of additions. The total number of steps needed is on the order of $\ell(N_1) \cdot \ell(N_2) < \ell(I)^2$.

Remark. This example illustrates that, when an n-tuple of inputs $(N_1,\ldots,N_n)$ is $\mathbf{N}$-encoded as $I = (N_1)_{(10)};(N_2)_{(10)};\ldots;(N_n)_{(10)}$ then a machine runs in $\tau(\ell(I))$ time for a polynomial $\tau(x)$ if and only if it runs in $\mu(\ell(N_1),\ldots,\ell(N_n))$ time for some polynomial of n variables μ. What if an algorithm requires, say, $N_1^2 \cdot \ell(N_2)^3$ steps? We wish to express the polynomial dependence on N_1 (not $\ell(N_1)$) and $\ell(N_2)$. To do this, we can redefine the problem so that the input consists of N_1 written in unary, followed by a semicolon and $(N_2)_{(10)}$. By "unary" we mean a string of N_1 1's to represent the number N_1, which we denote by $(N_1)_{(1)}$. With this redefinition we would have

$$I = (N_1)_{(1)};(N_2)_{(10)}$$

and consequently $\ell(I) = \ell((N_1)_{(1)}) + 1 + \ell((N_2)_{(10)}) = N_1 + 1 + \ell(N_2)$, hence $\ell(I)^5$ bounds the run time of $N_1^2\,\ell(N_2)^3$. Thus the problem "compute $f(N_1,N_2)$ from the input $I = (N_1)_{(1)}; (N_2)_{(10)}$" belongs to $\mathcal{P}$, even though the problem "compute $f(N_1,N_2)$ from the input $I = (N_1)_{(10)}; (N_2)_{(10)}$" does not. The only thing that has changed is the N-encoding used for the input. Because of this we will be very careful to spell out the N-encodings used when setting up a problem.

Definition 4.2

A function $f : \mathrm{Dom}(f) \to \mathbb{N}$, $\mathrm{Dom}(f) \subseteq \mathbb{N}$, belongs to the class $\underline{\#\mathcal{P}}$ if and only if there exists a nondeterministic Turing machine, together with a polynomial $\tau(x)$, with the following property. Whenever the input is $N \in \mathrm{Dom}(f)$, each of the possible paths taken by the machine reaches a terminal state within $\tau(\ell(N))$ steps. The number of possible paths which lead to the terminal state u_1 (as opposed to u_0) equals $f(N)$. Paths leading to the state u_1 are called _accepting_ paths.

Remark. It is important that $\tau(x)$ be universal with respect to the set of paths. That is, there is a single polynomial bound $\tau(\ell(N))$ such that any possible path uses fewer than this many steps.

Example. The function d, where $d(N)$ denotes the number of positive integral divisors of N, belongs to #$\mathcal{P}$. To see this, consider the following three-stage "nondeterministic program." First, write down any two numbers N_1 and N_2 of length $\leq \ell(N)$. To do this, the machine writes the first decimal digit, then the second decimal digit, and so on. Nondeterministic branching is involved at this point because, after writing a digit and comparing the current length with $\ell(N)$ and repositioning the head, it may enter any of $B_0 = 10$ states to get ready for the next digit. At the second stage multiply N_1 and N_2 to obtain a result N_3. Lastly, compare N and N_3; if equal enter state u_1, if unequal enter state u_0.

The ith stage of this program clearly takes at most $\tau_i(\ell(N))$ steps for some polynomials τ_i, so all paths terminate within $\tau(\ell(N))$ time if $\tau = \tau_1 + \tau_2 + \tau_3$. Furthermore, paths are classified by the pair (N_1, N_2) written during stage one, since the remainder of the program flows deterministically. The number of paths leading to state u_1 equals the number of pairs (N_1, N_2) such that $N_1N_2 = N$. Thus the number of accepting paths equals $d(N)$, as needed in Defintion 4.2.

Note. In the above argument it is implicitly assumed that "pairs of numbers written down during stage one" are in one-to-one correspondence with pairs of numbers (N_1, N_2) whose product N_3 is a candidate to equal N. This is correct except that leading zeros can create a problem; for instance, when $N = 30$ the two pairs (6,5) and (06,05) should not both be allowed. Stage one of the program must either forbid leading zeros or else require that N_1 and N_2 have length equal to exactly $\ell(N)$. This issue recurs when we are modeling a Turing machine via a vest in the proof of Theorem 4.5. In that proof we adopt the approach of insisting that the lead digit be nonzero.

Another example of a problem in $\#\mathcal{P}$ is the computation of the permanent of a square matrix whose entries lie in $\mathbb{N}$ [28]. The $\#\mathcal{P}$ problem with greatest relevance to the computational complexity of rational homotopy is described next.

Theorem 4.3

Computing the M-sequence of a vest is in $\#\mathcal{P}$. Specifically, let $f \subseteq \mathbb{N} \times \mathbb{N}$ be the function which takes as input the list of $1 + d + 1 + md^2 + 1 + hd$ integers (each written in decimal) describing a vest (d, v_0, T, S) followed by an integer n written in unary. The value of the function is the nth entry in the M-sequence of (d, v_0, T, S). Then f belongs to $\#\mathcal{P}$.

Remark. By the comments preceding Theorem 3.4, we know that a vest may be specified by a list of integers. Since these integers may be positive or negative or zero, however, it is useful to assume that a minus sign is included in the symbol set [B] available to our Turing machine.

Proof of Theorem 4.3. Consider the following program for a nondeterministic Turing machine. By successively writing down its decimal digits choose any index i satisfying $1 \leq i \leq m$. Then, (deterministically) compute T_iv_0 and replace the vector v_0 by the new value T_iv_0. Repeat these two steps exactly n times. Then (deterministically) compute Sv_0 and compare the result with the zero vector in $\mathbb{Z}^h$. If it is zero, enter state u_1; otherwise, enter state u_0.

The only nondeterministic part of the program concerns choosing the various indices, which we call i_1 through i_n. Permissible flow paths correspond to n-tuples $\sigma = (i_1, \ldots, i_n)$. A path ends in state u_1 if and only if it corresponds to a σ satisfying $S(v_0 * \sigma) = 0$. Thus the number of accepting paths equals e_n, the nth entry of the M-sequence for (d, v_0, T, S).

It remains to find a polynomial τ such that $\tau(\ell(I))$ bounds the run time regardless of the path taken. Let q denote the maximum absolute value of the inputs which are encoded in decimal. By induction on j, the entries of the vector $(T_{i_j} \cdots T_{i_1})(v_0)$ are bounded in absolute value by $d^{j-1}q^j$. During the jth pass through the loop we choose and perhaps copy one of m matrices, introducing bookkeeping on the order of $K_1md^2\ell(I)$ steps. We then multiply a $d \times d$ matrix, each of whose entries is bounded by q, by a $d \times 1$ vector, each of whose entries is bounded by d^nq^n. This requires d^2 multiplications and about d^2 additions, each of which needs at most $K_2 \log(d^nq^n) = K_2 n \log(dq)$ steps. Each pass through the loop uses on the order of $K_1md^2\ell(I) + K_3nd^2 \log(dq)$ steps, so the loop takes $K_1md^2n\ell(I) + K_3n^2d^2 \log(dq)$ steps. The final stage of the program, namely multiplication by S and comparison with zero, requires another $K_4hdn \log(dq)$ steps, so everything is certainly finished in $K_1md^2n\ell(I) + K_5nd(nd + h) \log(dq)$ steps. But notice that the numbers m, d, and h are all smaller than $\ell(I)$ because the input contains, if nothing else, $2 + d + md^2 + dh$ semicolons or separators. The number n is smaller than $\ell(I)$ because n is written in unary, and $\log(q) < \ell(I)$ because q is written in decimal. Thus all paths terminate in $K \cdot \ell(I)^5$ steps, as desired.

The function f described in Theorem 4.3 is actually a very special kind of #$\mathcal{P}$ problem. It is a "universal example" for the class #$\mathcal{P}$ in the sense that any $g \in \#\mathcal{P}$ can be factored through f, the other factor belonging to the class $\mathcal{P}$. This property, called "#$\mathcal{P}$-completeness," provides a kind of lower bound on the computational complexity of f. A precise definition follows.

Definition 4.4

Let $f : \mathrm{Dom}(f) \to \mathbb{N}$, $\mathrm{Dom}(f) \subseteq \mathbb{N}$, be any function. The function f is <u>#$\mathcal{P}$-complete</u> if $f \in \#\mathcal{P}$ and, for any $g \in \#\mathcal{P}$, there exists a function $\phi_g \in$

$\mathcal{P}$ such that, whenever $N \in \mathrm{Dom}(g)$, then $\phi_g(N) \in \mathrm{Dom}(f)$ and $f(\phi_g(N)) = g(N)$.

The function ϕ_g occurring in Definition 4.4 is called a "polynomial time many-one reduction," a phrase which happily we will use only once again. One may think of it as a preprocessor which quickly transforms or translates (in the sense of translating a language) the original input N into an $I = \phi_g(N)$ which is suitable for use by f. The point is that <u>any</u> $g \in \#\mathcal{P}$ can, modulo a "fast" translation, be viewed as a special case of f. In other words, a $\#\mathcal{P}$-complete function f is so powerful that it already encompasses, in thinly disguised form, <u>every</u> $\#\mathcal{P}$ problem. On top of that, it is itself in $\#\mathcal{P}$!

Remarkably, the class $\#\mathcal{P}$ does contain universal examples of this type. The first example to be discovered was the problem of computing the permanent of a matrix [28]. We claim next that the problem of computing an M-sequence also satisfies Definition 4.4.

Theorem 4.5

Computing the M-sequence of a vest is $\#\mathcal{P}$-complete. Specifically, the function f described in the statement of Theorem 4.3 is $\#\mathcal{P}$-complete.

<u>Proof</u>. Theorem 4.3 asserts that this f is in $\#\mathcal{P}$, so it remains only to verify the universal property. Let $g : \mathrm{Dom}(g) \to \mathbb{N}$ belong to $\#\mathcal{P}$. We must construct $\phi_g \in \mathcal{P}$ such that $g = (f \circ \phi_g)|_{\mathrm{Dom}(g)}$.

We really have very little to work with. The function g belongs to $\#\mathcal{P}$. Therefore there exists a nondeterministic Turing machine which, when given N Dom(g) as input, has all paths terminate in $\tau(\ell(N))$ steps, while the number of paths terminating in state u_1 equals $g(N)$. Let U, [B], Γ denote respectively the set of states, the symbol set, and the collection of transition quintuples for this Turing machine.

Here's the key idea. We will model the Turing machine $(U, [B], \Gamma)$ by a vest. The list of successive states experienced by the Turing machine will correspond to a sequence σ of linear transformations for the vest. A sequence σ which either (i) corresponds to a flow path which terminates in state u_0 or (ii) corresponds to an invalid flow path (i.e., some transitions not in Γ) will result in $S(v_0 * \sigma)$ being nonzero. On the other hand, a sequence σ which ends in state u_1 and which utilizes only valid transitions to get there will lead to $S(v_0 * \sigma)$ being zero. Thus the M-sequence entry e_n will be counting the number of valid paths ending in u_1, a number which equals $g(N)$.

The input I to the function f consists of a vest description along with a unary integer n. This N-encoded I is to be the output of the preprocessor ϕ_g. As it turns out, we may always take $d = 16$, and v_0 has fixed components except for two entries which equal N and N^2. So far, computing and writing down this much of the input takes $K_1 + K_2 \cdot \ell(N)^2$ time. The remainder of the vest description depends solely on g, so the algorithm for ϕ_g may "memorize" it and write it down in a constant number K_3 of steps. Lastly, the argument n may be taken to be any integer greater than $\tau(\ell(N)) \cdot (\tau(\ell(N)) + \ell(N) + 1)$. To compute this number and to write it down in unary also takes a polynomial in $\ell(N)$ steps. Thus ϕ_g runs in polynomial time.

We will now describe explicitly how a vest can model the nondeterministic Turing machine $(U,[B],\Gamma)$ with input N, so that the M-sequence entry e_n equals $g(N)$ when $n \gg 0$. As noted above, the dimension d of the vector space involved in the vest will always be $d = 16$. The number m of transformations will be $\#(T) = \#(\Gamma) + 1 + 4B$. And h will equal 1.

To describe the m transformations it is useful to borrow from computer science the concept of "registers." A register is a dedicated, easily accessed, named computer memory location with enough space to store a single integer. Each of the 16 coordinates of the vector v_0, before and after applying the various transformations, will be thought of as a register. A list of these registers and their purposes is given in Figure 2; we elaborate on this outline in the text as well. For the time being the reader should ignore the rows and columns of Figure 2 which are marked with a superscript "a."

The first two registers are called "1" and PD (for path detector). The register denoted "1" always contains the numerical value 1, and each linear transformation is to leave this component of v_0 unchanged. The initial value of PD is zero. At any given moment, the path detector measures whether or not the sequence of transformations applied so far represents a valid flow path according to the Turing machine's transition quintuples Γ.

REGISTER NAME	INITIAL VALUES: COMPONENTS OF v_0	$^{(a)}$ $\tilde{v}_0$	DESCRIPTION OR PURPOSE
1	1	1	Always contains the numerical value 1
PD	0	0	Path Detector: remains zero as long as a sequence of transformations models a valid flow through the Turing machine, otherwise becomes positive
R	0	0	Describes tape contents to the right of the head, viewed as a base B integer written in reverse
R^2	0	0	$(R^2) = (R)^2$
L	N	0	Describes tape contents to the left of the head, viewed as a base B integer
L^2	N^2	0	$(L^2) = (L)^2$
H	0	0	Describes the tape symbol directly under the head
H^2	0	0	$(H^2) = (H)^2$
F	2	2	Contains the code for the current state of the Turing machine (note: $\nu(u^*) = 2$)
F^2	4	4	$(F^2) = (F)^2$
A	1	-1	Used to help control the flow of the program. Codes are as follows: -1: first digit of SI 0: subsequent digits of SI, or like 1 1: ready for T_0 or a T_γ -2 (+2): first digit of M, head moving left (right) -3 (+3): subsequent digits of M, head moving left (right)
A^2	1	1	$(A^2) = (A)^2$
M	0	0	A work space in which the integer part of (R)/B or of (L)/B is constructed non-deterministically
M^2	0	0	$(M^2) = (M)^2$
MR	0	0	$(MR) = (M)\cdot(R)$
ML	0	0	$(ML) = (M)\cdot(L)$
(a) SI		0	Simulated Input: stores an arbitrary number constructed non-deterministically before the modeling of the Turing machine begins
(a) NS		0	Number of Steps: every matrix increments (NS) by one each time it acts

Figure 2 List of the 16 (resp. 18) registers, along with their initial values in v_0 (resp. $\tilde{v}_0$) and their interpretations.

[a]This information is relevant to Theorem 4.8 only.

In a vest any sequence $T_{i_1} \cdots T_{i_n}$ of matrices must be allowed, but only certain transitions are permitted in the Turing machine. To get around this problem we permit the use of any matrix at any stage, but record in the register PD whether or not the selected transformation represents a valid continuation of the path. If so, PD is unchanged; if not, it is increased. At the end, a positive value in PD signifies an invalid path. The $h \times d$ matrix S of the vest measures PD. A necessary condition for $S(v_0 * \sigma)$ to be zero will be that the PD-component of $v_0 * \sigma$ vanish.

The remaining 14 registers come in pairs, denoted $F,F^2;R,R^2;H,H^2;L,L^2;M,M^2;A,A^2;MR,ML$. The names are indeed suggestive: within each pair except the last, one register's contents will always be the square of the other's contents, and each transformation will act so as to preserve this relationship. The last two registers always contain the products of the contents of M and of R, and of M and of L, respectively. To express this in symbols, let (X) denote the contents of the register X. We are saying that the relations $(F^2) = (F)^2$, $(MR) = (M) \cdot (R)$, etc. always hold.

Let $\nu : U \to \mathsf{N}$ be any injection which assigns distinct integers to the internal states U. For convenience assume $\nu(u_0) = 0$, $\nu(u_1) = 1$, $\nu(u^*) = 2$. The contents of (F,F^2) will always be $(\nu(u), \nu(u)^2)$ when the Turing machine being modeled is in the state $u \in U$. In particular, the initial values (in terms of components of v_0) for (F,F^2) are $(\nu(u^*), \nu(u^*)^2) = (2,4)$.

The next three register pairs tell us what is on the tape, thus completing the description of the machine's configuration. Since the tape is infinite in both directions, with only finitely many squares nonblank, we can summarize its contents as three finite integers written in base B. The single symbol directly under the head gives the contents of H. The base B number found by reading that portion of the tape to the left of the head (its units digit is in the square immediately to the left of the head) gives the contents of L. Likewise, that portion of the tape to the right of the head can be interpreted as a base B number written in reverse (units digit immediately to the right of the head), and this number gives the contents of R. Because of our convention that the head starts out on the first blank square to the right of the input, our initial values for (L, L^2) are (N, N^2). (This is why ϕ_g had to compute and write down N and N^2.) The registers H, H^2, R, R^2 all have initial values of zero, indicating blank regions of the tape.

We describe next the $m = \#(\Gamma) + 1 + 4B$ matrices in the set T, and in the process we illuminate the roles of the remaining registers. The action of each of these linear transformations is summarized in Figure 3. Each $\gamma \in \Gamma$ is of course a quintuple, $\gamma = (\gamma_1, \gamma_2, \gamma_3, \gamma_4, \gamma_5)$, and to each $\gamma \in \Gamma$ we associate one matrix T_γ whose precise effect on Q^d is described in Figure 3. Basically, when a vector v has components which describe the configuration of our Turing machine, then $T_\gamma v$ will describe the configuration after the transition γ. (If γ represents a transition from a configuration other than that described by v, then the path detector will be incremented.)

We also stipulate an extra transitionlike matrix T_0, which has the effect of leaving all registers intact except PD, which is incremented unless the machine's state is u_1. Including T_0 simulates the idea of making u_1 an absorbing rather than a terminal state: if the Turing machine enters u_1, it keeps returning to u_1 forever. With this change, the number of accepting paths becomes recast as the number of paths which survive for more than $\tau(\ell(N))$ steps. Thus e_n, which for the M-sequence counts the number of valid matrix products of length n, equals precisely the number of accepting paths once n is large enough.

It seems that we have everything we need already, but a problem arises which motivates the inclusion of the remaining 4B matrices. If a transition γ did not move the head, there would be no difficulty representing the effect on the register vector v by a linear transformation. Now suppose the head is moved by γ, say one square to the right, a move signaled by $\gamma_5 = +1$. The new contents of L become $B \cdot (L) + (H)$, which is a linear change. But the new contents of H become the unit's digit (base B) of the old (R), and the new contents of R are the old (R) with its units digit truncated. Both of these are highly nonlinear! Thus no single linear transformation will achieve all of the desired effects on the register vector v. The transition γ cannot be simulated by a single matrix T_γ.

A specific instance may help to clarify this. Suppose the tape reads 12763514, with the head positioned over the six. Our registers will con-

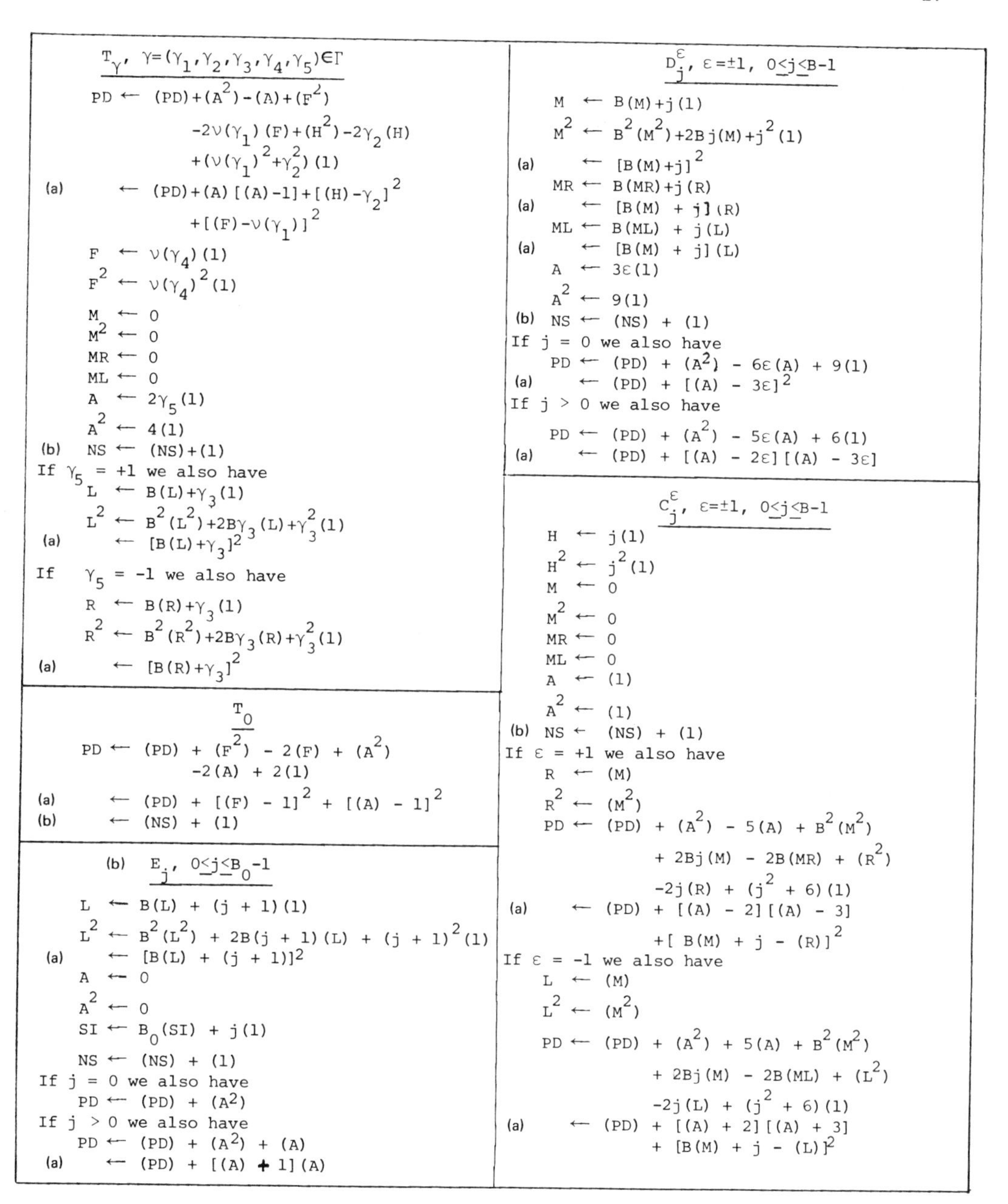

$T_\gamma,\ \gamma=(\gamma_1,\gamma_2,\gamma_3,\gamma_4,\gamma_5)\in\Gamma$

$PD \leftarrow (PD)+(A^2)-(A)+(F^2) -2\nu(\gamma_1)(F)+(H^2)-2\gamma_2(H) +(\nu(\gamma_1)^2+\gamma_2^2)(1)$

(a) $\leftarrow (PD)+(A)[(A)-1]+[(H)-\gamma_2]^2 +[(F)-\nu(\gamma_1)]^2$

$F \leftarrow \nu(\gamma_4)(1)$

$F^2 \leftarrow \nu(\gamma_4)^2(1)$

$M \leftarrow 0$

$M^2 \leftarrow 0$

$MR \leftarrow 0$

$ML \leftarrow 0$

$A \leftarrow 2\gamma_5(1)$

$A^2 \leftarrow 4(1)$

(b) $NS \leftarrow (NS)+(1)$

If $\gamma_5 = +1$ we also have

$L \leftarrow B(L)+\gamma_3(1)$

$L^2 \leftarrow B^2(L^2)+2B\gamma_3(L)+\gamma_3^2(1)$

(a) $\leftarrow [B(L)+\gamma_3]^2$

If $\gamma_5 = -1$ we also have

$R \leftarrow B(R)+\gamma_3(1)$

$R^2 \leftarrow B^2(R^2)+2B\gamma_3(R)+\gamma_3^2(1)$

(a) $\leftarrow [B(R)+\gamma_3]^2$

T_0

$PD \leftarrow (PD) + (F^2) - 2(F) + (A^2) -2(A) + 2(1)$

(a) $\leftarrow (PD) + [(F) - 1]^2 + [(A) - 1]^2$

(b) $\leftarrow (NS) + (1)$

(b) $E_j,\ 0\le j\le B_0-1$

$L \leftarrow B(L) + (j + 1)(1)$

$L^2 \leftarrow B^2(L^2) + 2B(j + 1)(L) + (j + 1)^2(1)$

(a) $\leftarrow [B(L) + (j + 1)]^2$

$A \leftarrow 0$

$A^2 \leftarrow 0$

$SI \leftarrow B_0(SI) + j(1)$

$NS \leftarrow (NS) + (1)$

If $j = 0$ we also have

$PD \leftarrow (PD) + (A^2)$

If $j > 0$ we also have

$PD \leftarrow (PD) + (A^2) + (A)$

(a) $\leftarrow (PD) + [(A) + 1](A)$

$D_j^\varepsilon,\ \varepsilon=\pm1,\ 0\le j\le B-1$

$M \leftarrow B(M)+j(1)$

$M^2 \leftarrow B^2(M^2)+2Bj(M)+j^2(1)$

(a) $\leftarrow [B(M)+j]^2$

$MR \leftarrow B(MR)+j(R)$

(a) $\leftarrow [B(M) + j](R)$

$ML \leftarrow B(ML) + j(L)$

(a) $\leftarrow [B(M) + j](L)$

$A \leftarrow 3\varepsilon(1)$

$A^2 \leftarrow 9(1)$

(b) $NS \leftarrow (NS) + (1)$

If $j = 0$ we also have

$PD \leftarrow (PD) + (A^2) - 6\varepsilon(A) + 9(1)$

(a) $\leftarrow (PD) + [(A) - 3\varepsilon]^2$

If $j > 0$ we also have

$PD \leftarrow (PD) + (A^2) - 5\varepsilon(A) + 6(1)$

(a) $\leftarrow (PD) + [(A) - 2\varepsilon][(A) - 3\varepsilon]$

$C_j^\varepsilon,\ \varepsilon=\pm1,\ 0\le j\le B-1$

$H \leftarrow j(1)$

$H^2 \leftarrow j^2(1)$

$M \leftarrow 0$

$M^2 \leftarrow 0$

$MR \leftarrow 0$

$ML \leftarrow 0$

$A \leftarrow (1)$

$A^2 \leftarrow (1)$

(b) $NS \leftarrow (NS) + (1)$

If $\varepsilon = +1$ we also have

$R \leftarrow (M)$

$R^2 \leftarrow (M^2)$

$PD \leftarrow (PD) + (A^2) - 5(A) + B^2(M^2) + 2Bj(M) - 2B(MR) + (R^2) -2j(R) + (j^2 + 6)(1)$

(a) $\leftarrow (PD) + [(A) - 2][(A) - 3] +[B(M) + j - (R)]^2$

If $\varepsilon = -1$ we also have

$L \leftarrow (M)$

$L^2 \leftarrow (M^2)$

$PD \leftarrow (PD) + (A^2) + 5(A) + B^2(M^2) + 2Bj(M) - 2B(ML) + (L^2) -2j(L) + (j^2 + 6)(1)$

(a) $\leftarrow (PD) + [(A) + 2][(A) + 3] + [B(M) + j - (L)]^2$

Figure 3 Explicit description of the action of each transformation on the registers.

Rules of the form $X \leftarrow (X)$ are omitted. If a register is not altered by a particular transformation, this information is omitted.

[a]Denotes a nonlinear equivalent expression (see the proof of Lemma 4.6, claim 2).

[b]Relevant to Theorem 4.8 only.

tain (L,H,R) = (127,6,4153). If the head moves to the right without altering any symbols the new values of (L,H,R) should be (1276,3,415). This describes a nonlinear function of the three variables.

In order to get around this problem we introduce 4B additional matrices, calling them $D_j^+, D_j^-, C_j^+, C_j^-$, $0 \leq j \leq B-1$. Mnemonically, D stands for "add another digit" and C stands for "compare with the desired result," with the + or − signaling head motion right or left. Here's the plan. Rather than try to obtain (R)'s last digit directly, we use a nondeterministic approach. We detour to construct a new arbitrary number (M), one base B digit at a time, and then simply compare the result with (R). We can keep track of (M)'s units digit and the integer part of (M)/B as we construct it. In the likely event that (M) $\neq$ (R) we simply augment the path detector to jettison this sequence. But if (M) = (R), we now have (R)'s units digit and the characteristic of (R)/B in our hands!

The use of (M) to denote the new arbitrary integer was no accident: it will indeed be built in the register M. The register pair (A, A^2) will serve to control the flow of the program among the T_γ's, $D_j^{\pm}$'s, and $C_j^{\pm}$'s.

We can now describe the $h \times d$ matrix S of the vest: it is the 1×16 row vector which computes $(PD) + (F^2) - 2(F) + (A^2) - 2(A) + 2(1)$. Because of the relations $(F^2) = (F)^2$ and $(A^2) = (A)^2$, this formula coincides with the nonlinear formula $(PD) + [(F) - 1]^2 + [(A) - 1]^2$. Since we always have $(PD) \geq 0$, the formula vanishes only under the simultaneous conditions (PD) = 0 and (F) = 1 and (A) = 1. The condition (PD) = 0 reflects a valid flow path; $(F) = 1 = \nu(u_1)$ indicates an accepting path; and (A) = 1 is the code for having successfully completed any last movements of the Turing machine head.

This completes an overview of the vest which models the Turing machine $(U, [B], \Gamma)$ with input N. We now explore in detail the sequences σ of matrices which lead to $S(v_0 * \sigma)$ being zero.

Suppose we have a path of length $q \leq \tau(\ell(N))$ through the Turing machine. By this we mean a list $\delta = \{\delta_i\}_{1 \leq i \leq q} \subseteq \Gamma$ of transition quintuples, together with a list of quadruples $(u_{(i)}, r_i, h_i, l_i)_{0 \leq i \leq q} \subseteq U \times \mathbb{N} \times [B] \times \mathbb{N}$, with the following coherencies. We have $\delta_i = (u_{(i-1)}, h_{i-1}, (\delta_i)_3, u_{(i)}, (\delta_i)_5)$ for each i, $1 \leq i \leq q$. If $(\delta_i)_5 = +1$ we require

$l_i = Bl_{i-1} + (\delta_i)_3$ and $r_{i-1} = Br_i + h_i$. If $(\delta_i)_5 = -1$ we expect $r_i = Br_{i-1} + (\delta_i)_3$ and $l_{i-1} = Bl_i + h_i$. Briefly, $u_{(i)}$ denotes the ith state of the machine for the path δ (with $u_{(0)}$ being u^*) and (r_i, h_i, l_i) describes the tape immediately after the ith transition. It should be clear that δ and (r_0, h_0, l_0) uniquely determine $(u_{(i)}, r_i, h_i, l_i)$ for $1 \leq i \leq q$. Let Δ_N (resp. $\Delta_N{}^a$) consist of all paths δ for the input (0,0,N) (resp. all such paths having $u_{(q)} = u_1$).

Given $\delta = \{\delta_i\}$, let $J_{(i)}$ denote the matrix sequence

$$J_{(i)} = D_{j_t}{}^{\varepsilon} D_{j_{t-1}}{}^{\varepsilon} \cdots D_{j_1}{}^{\varepsilon} C_{j_0}{}^{\varepsilon} \qquad [12]$$

where ε is the symbol + (resp. −) if $(\delta_i)_5 = +1$ (resp. −1) and $j_t \cdots j_0$ is the base B representation of r_i (resp. l_i). It is assumed that $j_t > 0$ if $t > 0$. Let Σ denote the set of all sequences of matrices in T. Define $\Phi' : \Delta_N \to \Sigma$ by

$$\Phi'(\delta) = T_{\delta_1} J_{(1)} T_{\delta_2} J_{(2)} \cdots T_{\delta_q} J_{(q)} \qquad [13]$$

if the path δ has length q. The function Φ' is obviously injective. We have been building up to

Lemma 4.6

A matrix sequence σ satisfies $S(v_0 * \sigma) = 0$ if an only if S has the form of $\Phi'(\delta)$ followed by a string of T_0's, for some $\delta \in \Delta_N{}^a$.

Proof. We first verify four easy claims. For any $\sigma \in \Sigma$, we assert

1. The value of the A-component of $v_0 * \sigma$ is always 0, ±1, ±2, or ±3.
2. The relations $(F^2) = (F)^2$, $(MR) = (M) \cdot (R)$, etc. hold for the components of $v_0 * \sigma$.
3. Acting via any of the transformations on $v_0 * \sigma$ cannot decrease (PD).
4. The value of (PD) in $v_0 * \sigma$ is nonnegative.

Claim 1 is trivial, 2 is by induction on the length of σ, 3 relies on 1 and 2, and 4 follows by induction using 3.

The reader is now referred to Figure 4, which gives a "flowchart" for the vest. We are only interested in sequences σ of matrices which lead to $S(v_0 * \sigma) = 0$. Figure 4 indicates that certain matrices may succeed others only when certain requirements are met. These conditions are "enforced" by the fact that PD is rendered positive if they are disobeyed. The reader is welcome to check this using the explicit actions given in Figure 3.

Now suppose a matrix sequence $\sigma \in \Sigma$ does have the form specified in the statement of Lemma 4.6. By induction on j, one verifies that the components (registers) of

$$v_0 * T_{\delta_1} J_{(1)} T_{\delta_2} J_{(2)} \cdots T_{\delta_j} J_{(j)}$$

are determined by $(F,R,H,L,A,M,PD) = (\nu(u_{(j)}), r_j, h_j, l_j, 1, 0, 0)$. Once j reaches $q = \text{length}(\delta)$, we have $(F) = (A) = 1$ and $(PD) = 0$; hence repeated applications of T_0 will not alter these values. Thus $S(v_0 * \sigma) = 0$.

Conversely, suppose $S(v_0 * \sigma) = 0$. We know that $(PD) = 0$ and $(F) = (A) = 1$ at the end of this sequence of transformations. Since (A) starts out at one and (PD) ends up at zero we must begin σ with a T_γ having $(\gamma)_1 = u^*$. Since T_0 can be applied only when $(F) = (A) = 1$ and no quintuple γ has $(\gamma)_1 = u_1$, any occurrence of T_0 in σ can be followed only by another T_0. Because $(A) = (F) = 1$ at the end, σ must end with T_0 or a $C_j^{\pm}$. Thus σ has the form

$$\sigma = T_{\delta_1} \tilde{J}_{(1)} T_{\delta_2} \tilde{J}_{(2)} \cdots T_{\delta_q} \tilde{J}_{(q)} T_0^{\,s}$$

where $s \geq 0$ and $\tilde{J}_{(i)}$ is a product of $D_j^{\pm}$'s and $C_j^{\pm}$'s. Furthermore, Figure 4 quickly shows that each $\tilde{J}_{(i)}$ must have the form

$$\tilde{J}_{(i)} = D_{j_t}^{\varepsilon} \cdots D_{j_1}^{\varepsilon} C_{j_0}^{\varepsilon} \qquad \text{where } \varepsilon = (\delta_i)_5$$

The remainder of the argument involves induction on i. Suppose that, for some $i \geq 1$, we know that $\{\delta_1, \ldots, \delta_{i-1}\} \in \Delta_N$ and also that $\tilde{J}_{(k)} = J_{(k)}$ for $k < i$. One can determine that the components (registers) of $v_0 * T_{\delta_1} \cdots J_{(i-1)}$ accurately reflect the configuration of the Turing machine after the transitions $\delta_1, \ldots, \delta_{i-1}$. Then δ_i must be a

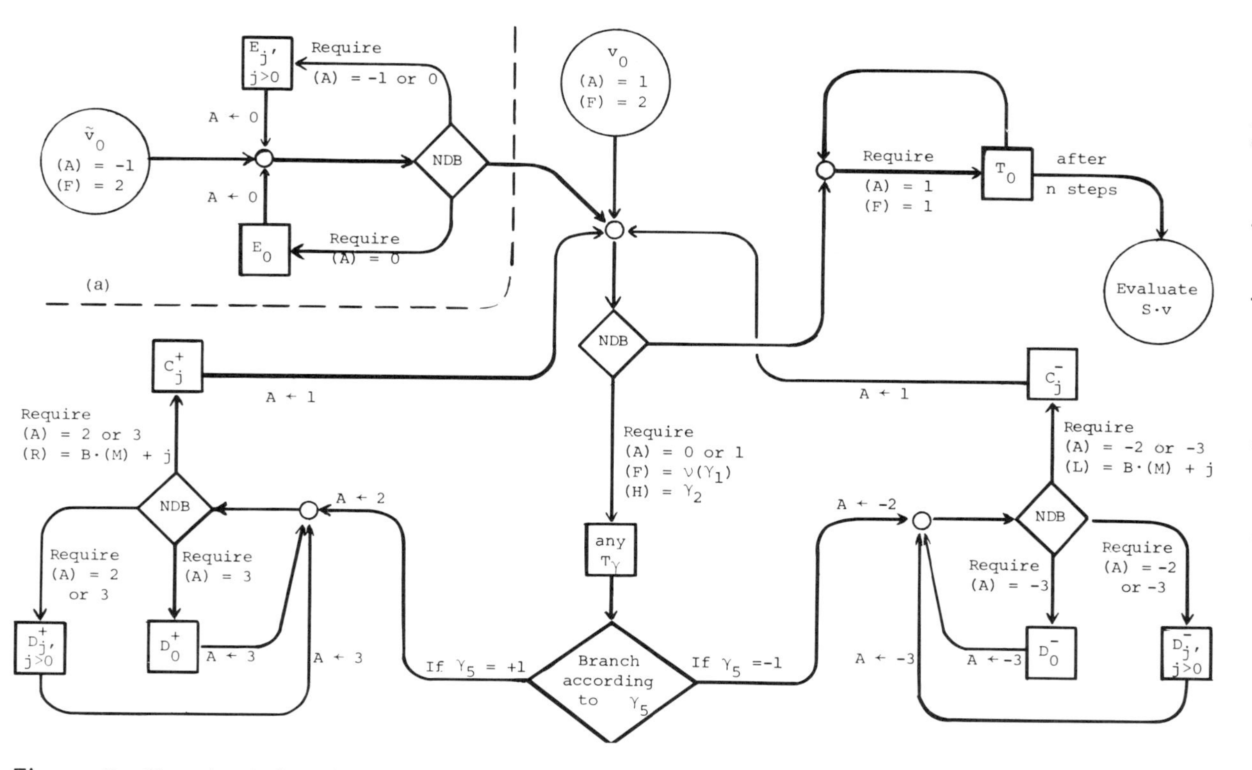

Figure 4 Flowchart for the vest which simulates a nondeterministic Turing machine.

"NDB" stands for "nondeterministic branching."

[a]The upper left region applies only to the vest of Theorem 4.8.

valid continuation of the path, and $\tilde{J}_{(i)}$ must equal $J_{(i)}$, or else (PD) would become positive. Eventually we see that $\delta = \{\delta_1, \ldots, \delta_q\} \in \Delta_N$. But we also know that $\nu(u_{(q)}) = 1$, so in fact $\delta \in \Delta_N{}^a$.

Lemma 4.7

Let $n \geq \tau(\ell(N)) \cdot (\tau(\ell(N)) + \ell(N) + 1)$. Sequences $\sigma \in \Sigma$ of matrices in T which have length n and which satisfy $S(v_0 * \sigma) = 0$ are in one-to-one correspondence with accepting paths on $(U,[B],\Gamma)$ for the input N.

Proof of Lemma 4.7. We observe first that any $\sigma \in \operatorname{im}(\Phi')$ has length $\leq n$. Since $\sigma = \Phi'(\delta)$ we have by (12) and (13) that s = length $(\sigma) \leq q(1 + p)$, where q is the path length of δ and p is the maximum length of any $J_{(j)}$. Recall that the length of $J_{(j)}$ equals the number of digits in the base B representation of either r_j or l_j. Note that the initial l_0 has $\ell(N)$ digits and that $r_0 = 0$. Since the Turing machine head can move just one square per step and it finishes in q steps, neither r_j nor l_j can ever exceed $\ell(N) + q$ digits. Thus $p \leq \ell(N) + q$ and

$$s \leq q(1 + q + \ell(N)) \leq \tau(\ell(N)) \cdot (\tau(\ell(N)) + \ell(N) + 1) \leq n$$

Now let $\Sigma_n{}^+$ consist of length n sequences σ for which $S(v_0 * \sigma) = 0$. Consider the function $\Phi : \Delta_N{}^a \to \Sigma$ given by, if $\Phi'(\delta)$ has length s, then $\Phi(\delta)$ consists of $\Phi'(\delta)$ followed by $(n - s)$ T_0's. By the previous paragraph this is well defined. Using Lemma 4.6, $\operatorname{im}(\Phi) = \Sigma_n{}^+$. Since Φ is injective, it offers the desired one-to-one correspondence.

It should now be clear that $e_n = \#\{$accepting paths for $(U,[B],\Gamma)$ with input $N\} = g(N)$. This completes the proof of Theorem 4.5.

We cannot resist proving one more theorem which uses the above construction. Theorem 4.8 opens up a vast array of sequences which can be shown to be M-sequences.

Theorem 4.8

Let $g : \mathbb{N} \to \mathbb{N}$ be any function whose domain is all of $\mathbb{N}$. Suppose that $g \quad \#\mathcal{P}$. More generally, suppose that there exists a nondeterministic Turing machine, together with constants $K > 0$ and $\varepsilon < \frac{1}{2}$, with the following property. When the input is N (given in base B_0), then the number of accepting paths equals g(N) and all paths terminate within $K \cdot N^\varepsilon$ steps. Then there is an M-sequence $\{\tilde{e}_n\}$ and an integer n_0 such that $\tilde{e}_n = g(n)$ for $n \geq n_0$.

<u>Remark</u>. The restriction that the domain be all of N is not a serious constraint in practice. For reasonable N-encodings one can generally decide deterministically in polynomial time (resp. in $K \cdot N^\varepsilon$ time) whether or not a given N belongs to Dom(g). If one extends such a $g : \mathrm{Dom}(g) \to \mathbf{N}$ over all of N by putting $\hat{g}|_{\mathrm{Dom}(g)} = g$ and $\hat{g}|_{\mathbf{N}-\mathrm{Dom}(g)} \equiv 0$, then $\hat{g}$ still belongs to #𝒫 (resp. satisfies the more general hypothesis of Theorem 4.8). The Turing machine for $\hat{g}$ is simply the machine for g together with a deterministic preprocessor which inspects the input and enters state u_0 if it lies in N − Dom(g).

<u>Proof of Theorem 4.8</u>. Our task is to construct a vest $(\tilde{d}, \tilde{v}_0, \tilde{T}, \tilde{S})$. We want valid sequences σ of length N ("valid" meaning $\tilde{S}(\tilde{v}_0 * \sigma) = 0$) to be in one-to-one correspondence with accepting paths on a Turing machine (U,[B],Γ) whose input is N.

We need to modify the vest developed for Theorem 4.5. For one thing, we need here a specific vest depending only on g. No dependence on an input is allowed, not even in the initial vector $\tilde{v}_0$. For another, we can no longer control the length. Sequences σ of arbitrary length must be allowed to occur.

Essentially, our new vest models a Turing machine which is (U,[B], Γ) preceded by a nondeterministic preprocessor. This preprocessor writes down a random base B_0 integer N, called the "simulated input," whose lead digit is nonzero. It does this by writing any digit and then moving to the right over the tape, repeating this step as often as it feels like it. After each such step it is also free to flip into state u^*. When it does eventually enter state u^*, it proceeds to view N as its input and to compute g(N) ("compute" in the nondeterministic sense of path counting).

In order to model this new machine, we keep the 16 registers from before and we introduce two new ones, bringing the total vector space dimension up to $\tilde{d} = d + 2 = 18$. The new registers are called SI for "simulated input" and NS for "number of steps." We also extend the set of matrices T to a larger set $\tilde{T}$ by introducing B_0 new matrices called E_j, $0 \leq j \leq B_0 - 1$. Lastly, we extend S from a $1 \times d$ to a $2 \times \tilde{d}$ matrix $\tilde{S}$, the second component evaluating (NS) − (SI). Unless both components of $\tilde{S}(v_0 * \sigma)$ are zero, the matrix sequence σ is rejected as invalid.

The NS register counts the number of matrices which have acted on $\tilde{v}_0$. Its initial value is zero, and every transfromation in $\tilde{T}$, including the E_j's, will increase (NS) by 1. Thus the NS component of $\tilde{v}_0 * \sigma$ will equal s if σ has length s.

The SI register is not affected by any of the T_γ's, $D_j^{\pm}$'s, $C_j^{\pm}$'s, or T_0. The initial value of (SI) in $\tilde{v}_0$ is zero. The effect of the E_j's is to build an arbitrary number N in the register L but simultaneously to store that number in the register SI. Once state u^* is entered, (L) will start to change but (SI) is not touched. The SI register maintains a permanent record of the simulated input.

[Recall that a number in base B_0 uses tape symbols 1 through B_0, not 0 through $B_0 - 1$. This explains why (j + 1) and not j gets involved when building the simulated input N in register L; see the action of E_j in Figure 3.]

The initial values for $\tilde{v}_0$ are listed in the third column of Figure 2. The fact that (R) = (L) = (H) = 0 simulates an initially blank tape. Having (A) = −1 forces a valid σ to begin with an E_j, $j > 0$.

Valid sequences σ are those for which (PD) = 0 and (A) = (F) = 1 and (SI) = (NS) = length(σ) at the end. The full Figure 4 illustrates the possibilities. As noted just above, any valid σ must start out with an E_j where $j > 0$, otherwise PD is immediately ruined. Using Figure 4 and Lemma 4.6, we see that a valid σ must have the form

$$\sigma = (E_{j_t} \cdots E_{j_0})\Phi'(\delta)T_0 \cdots T_0, \qquad j_t > 0 \tag{14}$$

for some accepting path $\delta \in \Delta_n{}^a$, where $(n)_{(B_0)} = j_t \cdots j_0$.

Conversely, any sequence of the form (14) will result in the first component of $\tilde{S}(\tilde{v}_0 * \sigma)$ vanishing. We therefore ask, which σ of the form (14) cause the second component of $\tilde{S}(\tilde{v}_0 * \sigma)$ to vanish as well?

Let n_0 be large enough so that

$$N > K \cdot (N)^{\varepsilon}[K \cdot (N)^{\varepsilon} + \ell(N) + 1] + \ell(N)$$

whenever $N \geq n_0$ (n_0 exists because $\varepsilon < \frac{1}{2}$). Suppose that the simulated input n, which equals the base B_0 number $j_t \cdots j_0$ of (14), is greater than n_0. By the reasoning of Lemma 4.7, the sequence

$$E_{j_t} \cdots E_{j_0} \Phi'(\delta)$$

has length bounded by $\ell(n) + K \cdot (n)^{\varepsilon}[K \cdot (n)^{\varepsilon} + \ell(n) + 1]$, which in turn is smaller than n when $n \geq n_0$. Thus when $n \geq n_0$ there are exactly g(n) sequences of the form (14) which have length n and which have $j_t \cdots j_0 = (n)_{B_0}$. These are precisely the valid sequences σ, for which (SI) must equal (NS) in $\tilde{v}_0 * \sigma$. We have shown that $\tilde{e}_n = \#\{\sigma \mid \tilde{S}(\tilde{v}_0 * \sigma) = 0\} = g(n)$ when $n \geq n_0$, as desired.

Remark. It is trivial that $\mathcal{P} \subseteq \#\mathcal{P}$. More generally, if an algorithm for g(n) runs deterministically in $\tau(n)$ steps, then at most $\tau(n) + \ell(g(n)) + 1 \leq 2\tau(n)$ steps are needed to "compute" g(n) in the nondeterministic sense. Once g(n) has been written down on the tape, with the head positioned over its rightmost digit, just one nondeterministic pass through the number g(n) is needed in order to get the number of accepting paths to equal g(n). Thus Theorem 4.8 also applies to any g which can be computed deterministically in $K \cdot (n)^{\varepsilon}$ time.

Example. Let g denote the characteristic function for the set of primes. That is, g(n) = 1 if n is prime but g(n) = 0 for composite n. In view of [2] or [21], g(n) can be computed deterministically in $K \cdot (n)^{\varepsilon}$ steps, where $\varepsilon < 0.15$. Assuming the extended Riemann hypothesis, one even has $g \in \mathcal{P}$! By Theorem 4.8 we deduce the existence of a vest whose M-sequence has $e_n = g(n)$ for all sufficiently large n.

Observation. Although the fast primality-testing algorithms have a certain appeal, in truth the naive algorithm of "look for factors by dividing n by every number smaller that $\sqrt{n}$" will also work in the previous example. This algorithm takes $K_1 \cdot (\sqrt{n}) \log^{c_1}(n)$ time, so it appears to violate the hypotheses of Theorem 4.8. However, in the proof it suffices that every accepting path δ correspond under Φ' to a matrix sequence σ of length $< n$. Referring to the proof of Lemma 4.7, the length of $\Phi'(\delta)$ is actually bounded by the product of q, the maximum time needed, and p, the maximum space needed. Since the naive algorithm needs only polynomial space, this product is $K_2 \cdot (\sqrt{n}) \log^{c_2}(n)$, which grows more slowly than n. Theorem 4.8 actually applies to any nondeterministically computable g(n) for which the product of the maximum run time $\tau(n)$ and the maximum space needed $\rho(n)$ grows more slowly than n.

Corollary 4.9

Let g satisfy the hypotheses of Theorem 4.8 or the more general hypotheses suggested by the previous sentence. There exists a 123H-algebra A (resp. a finite simply connected CW complex X) whose Hilbert series $H_A(z)$ (resp. $P_{\Omega X}(z)$) is rationally related to the series $\sum_{n=0}^{\infty} g(n)z^n$.

Proof. By Theorem 4.8 we have

$$\sum_{n=0}^{\infty} g(n)z^n = M(z) + p(z)$$

for some M-series M(z) and some polynomial p(z) of degree $<n_0$. Consequently, M(z) is rationally related to $\sum_{n=0}^{\infty} g(n)z^n$. As noted in Section 3, we can arrange for $H_A(z)$ and $P_{\Omega X}(z)$ to be rationally related to M(z).

Example. Let g again denote the characteristic function of the set of primes. There is a 123H-algebra A whose Hilbert series $H_A(z)$ is rationally related to

$$\sum_{p\ \text{prime}}^{\infty} z^p = z^2 + z^3 + z^5 + z^7 + z^{11} + z^{13} + \cdots$$

Remark. Because some computations can be done more quickly on a multitape Turing machine than on a one-tape machine, we remark that multitape machines can also be modeled by vests. Now we think of the head as fixed, with any one of the tapes moving one square left or right during each transition. For each new tape we create three new register pairs to record its (R), (H), and (L). We also permit the control register A to assume two more pairs of values per tape, and one introduces more $C^{\pm}$'s and $D^{\pm}$'s accordingly. Thus the machines referred to in Theorem 4.8 and Corollary 4.9 can be taken to be multitape machines.

5. SOME #$\mathcal{P}$-HARD PROBLEMS

In Section 5 we offer rigorous definitions for Turing reducibility (in polynomial time), Turing equivalence, and #$\mathcal{P}$-hardness. We demonstrate that 12 natural problems concerning Hilbert and Poincaré sequences are all equivalent to one another and that all are #$\mathcal{P}$-hard. Certain conventions are adopted in the course of N-encoding these problems, and we engage in considerable discussion of the rationale behind these conventions.

The concept of Turing reducibility was originally formulated in order to describe a hierarchy of functions (or of sets) in recursion theory. Consider two functions, $f:\mathrm{Dom}(f) \to \mathbb{N}$ and $g:\mathrm{Dom}(g) \to \mathbb{N}$. The output $g(N)$ may be difficult or impossible to compute directly, but g and f might be related in such a way that $g(N)$ can be computed quickly provided that $f(N_1),\ldots,f(N_s)$ are available. Here it is assumed that the arguments N_i can be computed quickly from N and from $f(N_1),\ldots,$ $f(N_{i-1})$. Thus the problem g has been "reduced" to the problem of determining certain f-values.

In practice, a computer programmer might write a subroutine to evaluate f( ) and call it during the routine which computes g( ). The theoretical computer science analog is called an oracle. An <u>oracle</u> is a special state of a hypothetical Turing machine which is entered after an argument N_i is N-encoded onto a section of the tape. Once in the oracle state, the machine replaces N_i by $f(N_i)$ on the tape, a process which is presumed to require only $\ell(N_i)$ steps. The machine then enters an ordinary state and resumes its calculations.

In practice, this requires a user-transparent computation. In the world of theory, the Turing machine is permitted to look up $f(N_i)$ in a hypothetical table of f-values and to copy the answer in just $\ell(N_i)$ steps. Since f might even be a nonrecursive function, presuming the availability of such a table is akin to expecting magic or divine intervention, hence the term "oracle."

Definition 5.1

Let $f:\mathrm{Dom}(f) \to \mathbb{N}$ and $g:\mathrm{Dom}(g) \to \mathbb{N}$ be two functions. We say that g is <u>Turing reducible</u> to f [<u>in polynomial time</u>], written $g \underset{T}{\leq} f$, if and only if there exists a deterministic Turing machine with the following property. It has an oracle which looks up f-values, and it computes g(I) in polynomial time. [As before, "polynomial time" means that the number of steps is bounded by a fixed polynomial in $\ell(I)$.]

Remark. The relation $\leq_T$ is readily seen to be reflexive and transitive. We may interpret "$g \leq_T f$" as "g is no harder than f" for the following reason. If g is Turing reducible to f, then f being computable in polynomial time would make g computable in polynomial time. Ease of computing f immediately translates into ease of computing g. On the other hand, g might be strictly easier to obtain than f, since there could be other algorithms for g which do not even use the oracle. Contrapositively, if g cannot be computed in polynomial time, then neither can f; in this sense f is "at least as hard as g."

Definition 5.2

Let $f: \text{Dom}(f) \to \mathbb{N}$ and $g: \text{Dom}(g) \to \mathbb{N}$ be two functions. We say that f and g are Turing equivalent [in polynomial time], denoted $f \approx_T g$, if and only if both $f \leq_T g$ and $g \leq_T f$.

Terminology. In this chapter we will always say "Turing reducible (resp. equivalent)" when we mean "Turing reducible (resp. equivalent) in polynomial time."

It should be clear that Turing equivalence is an equivalence relation on functions from subsets of $\mathbb{N}$ to $\mathbb{N}$. As a nontrivial example, note that any $\#\mathcal{P}$-problem is Turing reducible to any $\#\mathcal{P}$-complete problem. Consequently, all $\#\mathcal{P}$-complete problems lie in the same Turing equivalence class.

Definition 5.3

A function $f: \text{Dom}(f) \to \mathbb{N}$ is $\#\mathcal{P}$-hard if and only if $g \leq_T f$ for every $g \in \#\mathcal{P}$.

This definition says roughly that a $\#\mathcal{P}$-hard problem is as hard as or harder than anything in $\#\mathcal{P}$. Note that all $\#\mathcal{P}$-complete problems are $\#\mathcal{P}$-hard, but not conversely.

Lemma 5.4

(i) If $f_0 \underset{T}{\leq} f_1$ and f_0 is #𝒫-complete or #𝒫-hard, then f_1 is #𝒫-hard.

(ii) If $f_1 \underset{T}{\approx} f_2$ and f_1 is #𝒫-hard, then f_2 is #𝒫-hard.

Proof. This is a trivial exercise in using the definitions.

We next present a list of nine problems which will all be shown to be Turing equivalent to one another and #𝒫-hard.

(A) Find the nth term of the Tor-sequence of a 123H-algebra A.

(B) Find the nth term of the Hilbert sequence of a 123H-algebra A.

(C) Find the nth term of the Hilbert sequence of a 12H-algebra A.

(D) Find the nth term of the Hilbert sequence of a 12-algebra A.

(E) Find the nth term of the Hilbert sequence of a degree-one-generated finitely presented connected graded Q-algebra A.

(F) Find the nth entry of the Poincaré sequence of a commutative local Q-algebra $(R, \mathfrak{M})$ for which $\mathfrak{M}^3 = 0$.

(G) Find the nth entry in the Poincaré-Betti sequence of a finite-dimensional basic local Q-algebra R.

(H) Evaluate $\dim_Q(H_n(\Omega X; Q))$, where X is a simply connected finite CW complex having cells (other than the base point) in dimensions two and four only.

(I) Determine $\dim_Q(\pi_{n+1}(X) \otimes Q)$, where X is as described in (H) above.

Let us specify how the input is to be N-encoded for each of these problems. Problems (A) through (D) call for an integer n and a description of a 12-algebra. In the presentation

$$A = Q\langle x_1, \ldots, x_g\rangle / \langle \alpha_1, \ldots, \alpha_r\rangle$$

we have

$$\alpha_k = \sum_{i=1}^{g} \sum_{j=1}^{g} c_{ijk} x_i x_j \qquad [15]$$

for certain constants $c_{ijk} \in Q$. We can assume, by clearing denominators if necessary, that $c_{ijk} \in Z$. To specify a general 12-algebra A we will therefore give g and r in decimal, followed by the $g^2 r$ decimal integers $\{c_{ijk}\}$. The input n then follows in unary.

Objections. This N-encoding is appropriate for very complicated or "generic" presentations, but it seems very wasteful for relatively simple algebras, which tend to occur in practice. Consider the presentation

$$A = Q\langle x_1,\ldots,x_{10}\rangle / \langle x_1x_2 + 6x_3x_4, x_5x_6, x_7x_8, x_9x_{10}\rangle \qquad [16]$$

Our N-encoding will require over 400 entries, nearly all of them being zero, whereas (16) presents the same information in just one line. However, if concise inputs like (16) were sometimes allowed, this could skew the computational complexity of the overall problem and throw off the upcoming theorem on Turing equivalence.

The situation is analogous to that of scientific or other space-saving notations. Borrowing from FORTRAN the notation xEy for $10^y x$, note that the answer to the addition problem "2E1000 + 3E25" would have to be written as

20000···[974 zeros]···0003E25

The size of this output is not a polynomial in the input length, which suggests that addition does not belong to $\mathcal{P}$. To avoid this false or at best misleading conclusion we require that the input always be written out in full decimal notation, no matter how inefficient this seems for a given problem. Likewise, we always insist that the presentation for a 12-algebra be N-encoded as described above.

A second possible objection to our proposed N-encoding is that it is redundant for 12H-algebras. When A is a 12H-algebra the constants $\{c_{ijk}\}$ satisfy $c_{ijk} = c_{jik}$, so nearly half of them are superfluous. The response to this objection is that it doesn't matter, since cutting the input size by a factor of two has no effect on the computational complexity. To make this precise, consider a problem (C′), which seeks the same output as problem (C) but whose input omits the double entries. In polynomial time an N-encoded input for (C′) can be converted to the corresponding input for (C), and also vice versa, so (C′) and (C) are Turing equivalent. In general, when one has several choices of N-encoding for the input to a problem, its Turing equivalence class is independent of the choice as long as the various inputs can be obtained from one another in polynomial time.

Returning to our list of problems, the input for problem (E) is handled similarly. Now the algebra A is allowed to have a presentation of the form

$$A = Q\langle x_1,\ldots,x_g\rangle / \langle \beta_1,\ldots,\beta_r\rangle \qquad [17]$$

where each $|x_i| = 1$ but the relations' degrees $t_k = |\beta_k|$ may be arbitrary positive integers. To describe the β_k's let Λ_s denote the set of length s sequences whose entries come from $\{1,2,\ldots,g\}$, and for $\lambda = (i_1,\ldots,i_s) \in \Lambda_s$ let x_λ denote the monomial $x_{i_1}\cdots x_{i_s}$. A typical relation β_k has the form

$$\beta_k = \sum_{\lambda \in \Lambda_{t_k}} c_{\lambda k} x_\lambda, \qquad c_{\lambda k} \in Q$$

where again we may clear denominators and assume that $c_{\lambda k} \in Z$. In order to specify a degree-one-generated algebra A we write

$$(g)_{(10)};\ (r)_{(10)};\ (t_1)_{(10)};\ldots;\ (t_r)_{(10)}$$

followed by the $\{c_{\lambda k}\}$ in decimal. The input n is in unary.

Now consider problems (F) and (G). A <u>finite-dimensional basic local Q-algebra</u> is an associative ring R with unity such that $R/\mathcal{R}(R) \approx Q$ and $\dim_Q(R) < \infty$, where $\mathcal{R}(R)$ denotes the Jacobson radical of R. Viewing Q as an R-module, we may form the <u>Poincaré-Betti sequence</u> $\{\rho_n\}$, where

$$\rho_n = \dim_Q(\mathrm{Tor}_n^R(Q,Q))$$

The Poincaré sequence of a commutative Artinian Q-algebra $(R,\mathfrak{M})$ having $\mathfrak{M}^3 = 0$ is a very special case. The rings discussed in problem (F) are always isomorphic to polynomial ring quotients of the form

$$Q[y_1,\ldots,y_g]/J$$

Here J denotes an ideal satisfying $\underline{\mathfrak{a}}^2 \supseteq J \supseteq \underline{\mathfrak{a}}^3$ when $\underline{\mathfrak{a}}$ denotes the ideal $(y_1,\ldots,y_g)$.

A general finite-dimensional basic local Q-algebra R may be written as $Q \oplus \mathcal{R}(R)$. If $\{b_1,\ldots,b_s\}$ is a Q-basis for $\mathcal{R}(R)$, the multiplication on R determines s^3 constants d_{ijk} via

$$b_i b_j = \sum_{k=1}^{s} (d_{ijk}) b_k \qquad [18]$$

Conversely, the constants $\{d_{ijk}\} \subseteq Q$ determine R.

We may convert all the $\{d_{ijk}\}$ to whole numbers by choosing a least common denominator δ and by using $\{b_i'\}$ instead of $\{b_i\}$ for our basis, where $b_i' = \delta b_i$. This converts (18) to the expression

$$b_i' b_j' = \sum_{k=1}^{s} (d_{ijk}') b_k'$$

where $d_{ijk}' = \delta d_{ijk} \in Z$. Our N-encoding for problems (F) and (G) will consists of s and the s^3 integers $\{d_{ijk}'\}$, all written in decimal, together with n written in unary.

As to problems (H) and (I), the space X is determined up to homotopy type by the collection of r homotopy classes $[f_k] \in \pi_3(W)$, where $W = \vee_{i=1}^{g} S^2$. Since it is a free abelian group on $\frac{1}{2}g(g + 1)$ generators, $\pi_3(W)$ is easy to understand. When $\iota_1, \ldots, \iota_g$ denote the natural generators of $\pi_2(W) \approx Z^g$, then $\pi_3(W)$ is generated by the g compositions $\iota_i \circ \eta$ (η denotes the Hopf map) and by the $\frac{1}{2}g(g - 1)$ Whitehead products $[\iota_i, \iota_j]$ for $i < j$. The homotopy class $[f_k]$ may be written

$$[f_k] = \sum_{i=1}^{g} c_{iik}[\iota_i \circ \eta] + \sum_{i<j} c_{ijk}[\iota_i, \iota_j]$$

for some integers $\{c_{ijk}\}$. In order to specify X it suffices to list the $\frac{1}{2}r(g^2 + g)$ coefficients $\{c_{ijk} \mid 1 \leq i \leq j \leq g,\ 1 \leq k \leq r\}$. Our N-encoding will give r and g and the $\{c_{ijk}\}$ in decimal, followed by n in unary.

Proposition 5.5

Problem (A) is #$\mathcal{P}$-hard.

Proof. By Lemma 5.4(i) and Theorem 4.5, we need only show that the problem "compute the nth entry in the M-sequence of a vest" is Turing reducible to problem (A). This reduction is implicit in Theorem 3.4, as long as the 123H-algebra corresponding to a given vest can be N-encoded in polynomial time.

The proof in [4] of Theorem 3.4 indicates how this can be done. From the vest $(d, v_0, \{T_i\}, S)$ we get a 123H-algebra A having $g = 2m + d + h + 3$ generators and $r = (m + 1)(m + d + h + 2) + 1$ relations. Some of the constants $\{c_{ijk}\}$ in A's presentations are 0 or ±1 according to a straightforward pattern, and the remainder may simply be copied from the entries of the $\{T_i\}$ or S.

Remark. The reduction described in the above proof is actually of the special "polynomial time many-one" type mentioned in Section 4. If it were also true that problem (A) belonged to #$\mathcal{P}$, then it would be #$\mathcal{P}$-complete. However, the author doubts that (A) $\in$ #$\mathcal{P}$.

Theorem 5.6

Problems (A) through (I) are all Turing equivalent to one another, and all are #$\mathcal{P}$-hard.

Proof. In view of Lemma 5.4(ii) and Proposition 5.5, it suffices to demonstrate the Turing equivalence of the nine problems.

The calculations in Section 3 indicate the equivalences of (A) with (B) and of (H) with (I). For instance, if the answer to (I) were available through an oracle, one could make a list on the tape of the rational homotopy ranks $\{r_j(X)\}$ for $1 \leq j \leq n$ and then apply the indicated procedures to determine $\dim(H_n(\Omega X; Q))$. Notice how critical it is here that n be given in unary. Both the listing of $\{r_j(X)\}$ and the subsequent computation require a length of time whose dependence on n is a polynomial in n. Because n is in unary, a polynomial in n is bounded by a polynomial in the input length.

Likewise, one easily obtains the equivalences (C) $\underset{T}{\approx}$ (H) and (see [6]) (C) $\underset{T}{\approx}$ (F). In the Turing reduction, the coefficients $\{c_{ijk}\}$ need only be copied and rearranged in preparation for the oracle. For instance, to show that (C) $\underset{T}{\leq}$ (H), we first convert the input I describing a 12H-algebra A into an input I' describing the associated four-dimensional CW complex X. We then invoke the oracle n times in order to list on our tape the first n terms of the Betti sequence of ΩX. Lastly, we calculate, as described in Section 3, the nth term of the Hilbert sequence for A. Each of these steps requires only a polynomial in $\ell(I)$ time.

It is trivial that (F) $\underset{T}{\leq}$ (G) and that (B) $\underset{T}{\leq}$ (C) $\underset{T}{\leq}$ (D). That (D) $\underset{T}{\leq}$ (B) is a consequence of [17, theorem 1.1]. Jacobsson's construction of a 123H-algebra whose Hilbert series is rationally related to that of an arbitrary 12-algebra takes only polynomial time.

We demonstrate next that (D) $\underset{T}{\approx}$ (E). The reduction (D) $\underset{T}{\leq}$ (E) is trivial. To show that (E) $\underset{T}{\leq}$ (D), use "link (c)-(d)" of [6]. In that article a construction is given which obtains from an arbitrary degree-one-generated A a 12-algebra A' such that $H_A(z)$ and $H_{A'}(z)$ are rationally related. When A has the presentation (17), then A' will have

$$g' = 1 + 3g + 9g^2 + \cdots + (3g)^{\bar{t}-1}$$

generators, where $\bar{t} = \max\{|\beta_1|,\ldots,|\beta_r|\}$. The coefficients c_{ijk} involved in the presentation of A' are 0 or ±1 or are obtained by copying the $c_{\lambda k}$'s. Using the facts that $g \geq 2$ (since the case $g = 1$ is trivial) and

$$g^{\bar{t}} = \#(\Lambda_{\bar{t}}) = \#\{c_{\lambda k'}\} < \ell(I)$$

when k' denotes an index such that $|\beta_{k'}| = \bar{t}$, we obtain

$$g' < (3g)^{\bar{t}} < g^{3\bar{t}} = (g^{\bar{t}})^3 < \ell(I)^3$$

Therefore the input I' describing the 12-algebra A' can be written down from the I which describes A, in an amount of time which is a polynomial in $\ell(I)$. It follows that (E) $\underset{T}{\leq}$ (D).

The above arguments show that eight of the nine problems, all except (G), are Turing equivalent. It remains only to show that (G) is reducible to one of the others; we will see that (G) $\underset{T}{\leq}$ (D). Applying the cobar construction to a finite-dimensional basic local Q-algebra R yields a free finitely generated differential graded algebra (B,d) for which

$$H^*(B,d) \approx \mathrm{Tor}^R_*(Q,Q)$$

as graded vector spaces. By Gulliksen's construction (see corollary 2 of [6]) there is a 12-algebra A whose Hilbert series is rationally related to the "homology series" of (B,d), which coincides with the Poincaré-

Betti series for R. Obtaining A's presentation from R's takes only polynomial time; again, it is mostly a reshuffling of the coefficients. This completes the proof of Theorem 5.6.

Remark. Notably absent from the list (A) through (I) is the problem of finding the Poincaré series of an arbitrary commutative Noetherian (not necessarily Artinian) local Q-algebra. There are two known reductions from such rings to problems in the list, namely Levin's reduction to Artinian rings [19] and "link (e)-(f)" of [6] to 12-algebras. However, neither reduction is certifiably accomplished in polynomial time. For instance, Levin's reduction requires us to choose a "sufficiently large" integer q so that $P_R(z)$ and $P_{R/(\mathfrak{M}^q)}(z)$ will be rationally related. The proof is nonconstructive and relies on the Artin-Rees lemma. If R is described via a presentation, e.g., as the quotient of a polynomial ring, the integer q need not be bounded by a polynomial in the size of the presentation. That even the length of q might grow faster than a polynomial in the presentation size is suggested by the results of [8], where some similar problems are studied.

We consider next three problems which are natural generalizations of problems (E), (H), and (I). We show that they too are Turing equivalent to the problems of Theorem 5.6, but we must make an unexpected assumption about their N-encodings.

(J) Find the nth term of the Hilbert sequence of a finitely presented connected graded Q-algebra A.

(K) Evaluate $\dim_Q(H_n(\Omega X;Q))$ for a finite simply connected CW complex X.

(L) Determine $\dim_Q(\pi_{n+1}(X) \otimes Q)$, where X is a finite simply connected CW complex.

Again we must be careful to specify how the input is to be N-encoded. As to problem (J), a presentation for the algebra A looks like

$$A = Q\langle y_1,\ldots,y_g\rangle/\langle\beta_1,\ldots,\beta_r\rangle \qquad [19]$$

where y_i has degree $m_i \geq 1$ and β_k has some homogeneous degree t_k. For a sequence $\lambda = (i_1,\ldots,i_q)$ of indices, $1 \leq i_j \leq g$, let $|\lambda|$ denote $m_{i_1} + \cdots + m_{i_g}$ and let y_λ denote the monomial $y_{i_1}\cdots y_{i_q}$ in the free associative algebra $F = Q\langle y_1,\ldots,y_g\rangle$. If $\tilde{\Lambda}_s$ denotes the set of sequences λ for which $|\lambda| = s$, then $\{y_\lambda \mid \lambda \in \tilde{\Lambda}_s\}$ is a Q-basis for the sth homogeneous piece F_s of F. Since each relation β_k is homogeneous of degree t_k, we can write (after clearing denominators)

$$\beta_k = \sum_{\lambda \in \tilde{\Lambda}_{t_k}} e_{\lambda k} y_\lambda, \qquad e_{\lambda k} \in Z$$

Thus the algebra A is specified by the "presentation degree vector"

$$(g; m_1, \ldots, m_g; r; t_1, \ldots, t_r)$$

and by the collection $\{e_{\lambda k}\}$ of coefficients. Let $\bar{t}$ denote $\max(t_1, \ldots, t_r)$.

To be consistent with problem (E) our input should provide

$$(g)_{(10)}; (m_1)_{(10)}; \ldots; (m_g)_{(10)}; (r)_{(10)}; (t_1)_{(10)}; \ldots; (t_r)_{(10)} \qquad [20]$$

followed by the $\{e_{\lambda k}\}$ in decimal and n in unary. However, with this N-encoding we run into difficulties. Let us denote by (J') the problem which seeks $\dim(A_n)$ from the input just described. It is trivial that (E) $\underset{T}{\leq}$ (J'); what happens if we try to prove that (J') $\underset{T}{\leq}$ (E)?

We must relate an arbitrary Hilbert series to that of a degree-one-generated algebra. One efficient way to do this is as follows. Starting with A given as in (19), let

$$\psi : Q\langle y_1, \ldots, y_g\rangle \to Q\langle x_0, x_1, \ldots, x_g\rangle$$

be the graded algebra homomorphism in which $|x_j| = 1$ and $\psi(y_i) = x_0^{m_i - 1} x_i$. Then ψ induces a monomorphism $\hat{\psi} : A \to \hat{A}$, where

$$\hat{A} = Q\langle x_0, x_1, \ldots, x_g\rangle / \langle \psi(\beta_1), \ldots, \psi(\beta_r)\rangle$$

is degree-one generated. By [5, theorems 3.1 and 2.4] one has the rational relationship

$$H_{\hat{A}}(z)^{-1} = H_A(z)^{-1} - (g + 1)z + (z^{m_1} + \cdots + z^{m_g})$$

The presentation for $\hat{A}$, which has as relations $\hat{\beta}_1 = \psi(\beta_1), \ldots, \hat{\beta}_r = \psi(\beta_r)$, can be obtained easily from knowledge of $\beta_1, \ldots, \beta_r$. If

$$\beta_k = \sum_{\lambda \in \tilde{\Lambda}_{t_k}} e_{\lambda k} y_\lambda$$

then

$$\hat{\beta}_k = \sum_{\lambda \in \tilde{\Lambda}_{t_k}} e_{\lambda k} x_{\tilde{\psi}(\lambda)} \qquad [21]$$

where $\tilde{\psi}: \tilde{\Lambda}_s \to \Lambda_s$ is the injection such that $\psi(y_\lambda) = x_{\tilde{\psi}(\lambda)}$. It appears that $\hat{\beta}_k$ can quickly be obtained from β_k. But in keeping with our previous N-encoding conventions, (21) should really be written out in full as

$$\hat{\beta}_k = \sum_{\sigma \in \Lambda_{t_k}} c_{\sigma k} x_\sigma \qquad [22]$$

where

$$c_{\sigma k} = \begin{cases} e_{\lambda k} & \text{if } \sigma = \tilde{\psi}(\lambda) \text{ for some } \lambda \in \tilde{\Lambda}_{t_k} \\ 0 & \text{if } \sigma \in \Lambda_{t_k} - \text{im}(\tilde{\psi}) \end{cases}$$

Now we see the problem which arises in passing from an input I' describing A to an input $\hat{I}$ for $\hat{A}$. Because $\#(\Lambda_{t_k})$ may grow much faster than $\#(\tilde{\Lambda}_{t_k})$, the number of coefficients which are in zero in (22) cannot be bounded by any fixed polynomial in $\ell(I')$. The reduction from (J') to (E) fails to take polynomial time, simply because there may be more than polynomially many zeros to write down! We must do one of three things: revise our N-encoding scheme for problem (E) so that it omits copious zeros; alter our N-encoding for problem (J'); or find another reduction, which does take polynomial time, from (J') to (E). The first course of action would again leave us open to the various objections discussed earlier, and I have not succeeded at the third, so we will adopt the second option.

In problem (J), the input I will consist of expression (20), the $\{e_{\lambda k}\}$ in decimal, $(g^{\bar{t}})_{(1)}$, and $(n)_{(1)}$. The insertion of $(g^{\bar{t}})_{(1)}$ guarantees that $\ell(I)$ will always exceed $g^{\bar{t}}$. The number of new zeros to be listed while describing the $\hat{\beta}_k$'s is smaller than

$$\#(\Lambda_{t_1}) + \#(\Lambda_{t_2}) + \cdots + \#(\Lambda_{t_r}) \leq r\, g^{\bar{t}} < \ell(I)^2$$

so now the reduction from (J) to (E) does take polynomial time. Thus (J) $\leq_T$ (E). Conversely, the input to (E) describing a degree-one-generated algebra already lists at least $g^{\bar{t}}$ coefficients, so its length exceeds $g^{\bar{t}}$ and only polynomial time is needed in order to insert $(g^{\bar{t}})_{(1)}$ into the input. Hence (E) $\leq_T$ (J) as well.

With the above N-encoding convention, we have shown

Lemma 5.7

Problems (E) and (J) are Turing equivalent.

Finally, let us turn our attention to problems (K) and (L). We will at last justify our title assertion that problem (L) is #$\mathcal{P}$-hard.

The input for (K) and for (L) is an N-encoded finite description of a finite simply connected CW complex X. We mentioned in Section 3 that we would rely upon the Quillen model of X for this description. The Quillen model is a free differential graded Lie algebra $(\mathcal{L}X, d_X)$ whose generators correspond with, but lie in one degree lower than, a basis for $\overline{H}_*(X;Q)$. Calling this basis $\{a_1, \ldots, a_h\}$ and letting $m_1, \ldots, m_h$ denote their degrees in $H_*(X;Q)$, we must specify their boundaries

$$d_X(a_k) \in (\mathcal{L}X)_{m_k - 2} = (L_Q\langle a_1, \ldots, a_h\rangle)_{m_k - 2}$$

Here $L_Q\langle S\rangle$ denotes the free Lie Q-algebra on a graded set S and $(L)_s$ denotes the degree s component of a graded group L.

For $x_1, \ldots, x_q \in \mathcal{L}X$ let $[x_1, \ldots, x_q]$ denote the repeated bracket

$$[\cdots[[x_1, x_2], x_3], \ldots, x_q]$$

Let Λ'_s consist of all sequences $\lambda = (i_1, \ldots, i_q)$ for which $(m_{i_1} - 1) + \cdots + (m_{i_q} - 1) = s$, and for each such λ let $[a_\lambda]$ denote $[a_{i_1}, \ldots, a_{i_q}]$. Any element y of $(\mathcal{L}X)_s$ may be written (not uniquely) as

$$y = \sum_{\lambda \in \Lambda'_s} e_\lambda [a_\lambda], \quad e_\lambda \in Q$$

In particular, $(\mathcal{L}X, d_X)$ is specified by h and the degree list $(m_1, \ldots, m_h)$, together with a list of coefficients $\{e_{\lambda k} \mid 1 \leq k \leq h,\ \lambda \in \Lambda'_{m_k - 2}\}$ for which

$$d_X(a_k) = \sum_{\lambda \in \Lambda'_{m-2}} e_{\lambda k}[a_\lambda]$$

We may always assume the $\{e_{\lambda k}\}$ to be whole numbers after replacing the $\{a_k\}$ by suitable multiples of themselves.

Let $\bar{m}$ denote $\max\{m_1, \ldots, m_h\}$. The input I for problems (K) and (L) will consist of the expression

$$(h)_{(10)};(m_1)_{(10)};\ldots;(m_h)_{(10)}$$

followed by the $\{e_{\lambda k}\}$ in decimal, followed by $(h^{\bar{m}})_{(1)}$ and $(n)_{(1)}$. By including $(h^{\bar{m}})_{(1)}$ in the input we guarantee that $\ell(I) > h^{\bar{m}}$, which is useful to us in much the same way that including $(g^{\bar{t}})_{(1)}$ in the input for problem (J) was useful.

Theorem 5.8

The 12 problems (A) through (L) are all Turing equivalent to one another, and all are #$\mathcal{P}$-hard.

Proof. In view of Theorem 5.6 and Lemma 5.7, it suffices to show that (H) $\underset{T}{\leq}$ (K), that (K) $\underset{T}{\leq}$ (J), and that (K) $\underset{T}{\approx}$ (L). The last of these claims follows from the remarks of Section 3 and the fact that n is given in unary.

That (H) $\underset{T}{\leq}$ (K) is virtually trivial. We need to know that, in polynomial time, a description $\hat{I}$ of a space X of dimension four can be converted into a description I of the same X as a general space. The coefficients c_{ijk} for $i < j$ (resp. $i = j$) in the input to (H) can simply be copied (resp. halved) to obtain suitable $e_{\lambda k}$'s. The only major difference between I and $\hat{I}$ is therefore the insertion of $(h^{\bar{m}})_{(1)}$ into $\hat{I}$. But here we have $\bar{m} \leq 4$, hence

$$\ell(h^{\bar{m}})_{(1)} = h^{\bar{m}} \leq h^4 = (g + r)^4 < \ell(I)^4$$

so only polynomial time is needed in order to build $\hat{I}$ from I.

For the Turing reduction (K) $\underset{T}{\leq}$ (J), use the Adams-Hilton model together with "link (a)-(b)" of [6]. By [7] the Adams-Hilton model for X may be taken to be the universal enveloping algebra of $(\mathcal{L}X, d_X)$, which may be written out in full in polynomial time. In the subsequent reduction to a finitely presented graded algebra, a space X whose reduced rational homology had a basis in dimensions $(m_1,\ldots,m_h)$ is associated to an algebra A having $g = h + 3$ generators in degrees

$$(1,1,2,m_1 - 1,m_2 - 1,\ldots,m_h - 1)$$

The relations for A occur in degrees no greater than $\bar{m} + 1$ and they may be written down in a straightforward manner from knowledge of the Adams-Hilton model. The N-encoded description for A has length which is a polynomial in $(h + 3)^{\bar{m}+1}$; this is bounded by a polynomial in $h^{\bar{m}}$, as needed. The proof is now complete.

Remark. The reader who is still concerned about the inclusion of $(g^{\bar{t}})_{(1)}$ (resp. $(h^{\bar{m}})_{(1)}$) in our input is reminded that this can only decrease the computational complexity. In other words, if (J') (resp. (K') or (L')) is the problem which seeks the same output as (J) (resp. (K) or (L)) but without the redundant unary in the input, then (J) $\leq_T$ (J') (resp. (K) $\leq_T$ (K'), (L) $\leq_T$ (L')). In view of Lemma 5.4 (i) and Theorem 5.8, we know that (J') (and (K') and (L')) are $\#\mathcal{P}$-hard. Thus we may assert without equivocation that "the computation of rational homotopy groups is $\#\mathcal{P}$-hard."

6. COMPUTABILITY IN EXPONENTIAL TIME

We will define "exponential time" and prove that problems (A) through (L) of the previous section can be computed in exponential time. This will complete our proof of the information summarized in Figure 1. We close with some philosophical remarks about computational complexity.

Definition 6.1

A function $f: \mathrm{Dom}(f) \to \mathbb{N}$, $\mathrm{Dom}(f) \subseteq \mathbb{N}$, is computable in exponential time if and only if there exists a deterministic Turing machine, together with a polynomial $\mu(x)$, having the following property. When the input is $N \in \mathrm{Dom}(f)$, the output is $f(N)$, and the total number of steps needed is bounded above by $2^{\mu(\ell(N))}$.

Lemma 6.2

If $f \underset{T}{\leq} g$ and g is computable in exponential time, then f is computable in exponential time.

Proof. If $f \underset{T}{\leq} g$, then f can be computed deterministically in $\tau(\ell(N))$ steps, $\tau(x)$ being a polynomial, provided we have access to an oracle for g. In particular, the oracle can be invoked at most $\tau(\ell(N))$ times, and the arguments N_i for which we request $g(N_i)$ all have length bounded by $\tau(\ell(N))$.

Modify the machine for f by replacing the special oracle state with a "subroutine call" to the machine which computes g in $2^{\mu(\ell(N_i))}$ time. The new machine still computes f and it runs in fewer than

$$\tau(\ell(N)) + (\tau(\ell(N))) \cdot (2^{\mu(\tau(\ell(N)))}) \qquad [23]$$

steps. Putting $\mu'(x) = \mu(\tau(x)) + \tau(x)$, we have expression (23) being majorized by $2^{\mu'(\ell(N))}$, as needed.

Lemma 6.3

If $g \in$ #𝒫, then g can be computed in exponential time.

Proof. In view of Lemma 6.2, it suffices to prove this for a single #𝒫-complete problem. Let g denote the problem of Theorems 4.3 and 4.5. The function g can be computed deterministically by serially considering each length n sequence $\sigma = (\sigma_1, \ldots, \sigma_n)$ and evaluating $v_0 * \sigma$. Each evaluation takes $\tau(\ell(I))$ steps, $\tau(x)$ being a polynomial, and there are $m^n < (\ell(I))^{\ell(I)}$ sequences to consider. The entire deterministic process is completed in

$$\tau(\ell(I)) \cdot \ell(I)^{\ell(I)} < 2^{\tau(\ell(I))+\ell(I)^2}$$

steps.

Theorem 6.4

Problem (D) of Section 5 can be computed in exponential time.

Proof. Let

$$A = Q\langle x_1,\ldots,x_g\rangle / \langle \alpha_1,\ldots,\alpha_r\rangle$$

be a 12-algebra. Let F denote the free algebra $F = Q\langle x_1,\ldots,x_g\rangle$ and let J denote the two-sided ideal in F generated by the relations $\alpha_1,\ldots,\alpha_r$, so that $A = F/J$. We view J as a graded vector subspace of $F = \oplus_{m=0}^{\infty} F_m$ and write $J = \oplus_{m=2}^{\infty} J_m$. The nth entry in the Hilbert sequence for A is

$$\dim(A_n) = \dim(F_n) - \dim(J_n) = g^n - \dim(J_n)$$

so we need only show that $\dim(J_n)$ can be computed in exponential time. Let I denote the input to problem (D) so that I N-encodes the algebra A and the integer n.

As a vector space, J_n is spanned by the finite set $\{u\alpha_k v\}$ as k runs from 1 to r and as (u,v) runs through all pairs of monomials in the $\{x_i\}$ for which $|u| + |v| = n - 2$. This finite set has cardinality

$$s = (n - 1)rg^{n-2} < (\ell(I)^2)(\ell(I))^{\ell(I)} < 2^{\ell(I)^2} \qquad [24]$$

From the expressions (15) we obtain s expressions of the form

$$u\alpha_k v = \sum_{i=1}^{g} \sum_{j=1}^{g} c_{ijk} u x_i x_j v \qquad [25]$$

Using (25), we may construct an $s \times g^n$ matrix C. Its rows are indexed by triples (u,k,v), its columns are indexed by the set of length n monomials w in F, and its entries are either c_{ijk} (if $w = ux_ix_jv$) or zero (if not). The row space of C is easily identified with J_n. We need only compute the rank of C.

Since C is an $s \times g^n$ matrix and both s and g^n are bounded by $2^{\ell(I)^2}$, we can write down C in exponential time. To find the rank of C, use Gaussian elimination. This requires fewer than $(2^{\ell(I)^2})^3 = 2^{3\ell(I)^2}$ operations. Our computation therefore requires only exponential time.

Combining this result with Lemma 6.2 and Theorem 5.8, we have at once

Corollary 6.5

Each of the 12 problems (A) through (L) described in Section 5 is computable in exponential time.

Combining this corollary with Lemma 6.3, we have these problems nicely bracketed between "#$\mathcal{P}$-hard" and "computable in exponential time."

Philosophical Remarks. What does it all mean? Rational homotopy groups can be computed in exponential time. But exponential time is generally considered too slow for practical implementation.

It is virtually certain that rational homotopy cannot be computed in polynomial time. If it could, then $\mathcal{P}$ would equal #$\mathcal{P}$, contradicting a widely believed conjecture. I believe further that problems (A) through (L) are strictly harder than anything in #$\mathcal{P}$. That is to say, problem (L) is (I believe) not Turing reducible to anything in #$\mathcal{P}$. Even if $\mathcal{P}$ were to equal #$\mathcal{P}$, rational homotopy might still be uncomputable in polynomial time.

Experience tells us that integral homotopy groups are considerably less accessible than rational homotopy groups, so integral homotopy could require exponential time or more. One runs into difficulty just trying to set up the problem of computing integral homotopy. We are thrown back upon the issue of how one N-encodes, with even moderate efficiency, an arbitrary simply connected finite CW complex. One typically needs more information than is available in the Quillen model. [A notable exception occurs for the subclass of spaces having cells in dimensions two and four only. These spaces are determined up to homotopy type by the input to problem (H).] Nevertheless, it is safe for practical purposes to declare that the problem of computing the whole $\pi_n(X)$ is #$\mathcal{P}$-hard.

Interestingly, if one considers a fixed simply connected Y and asks for $\pi_n(Y)$ as a function of n only, Curtis [11,12] showed that $\pi_n(Y)$ can be obtained using a semisimplicial group model for Y whose lower central series has been truncated at depth 2^n. This suggests that for each Y, the function $\theta(n) = \pi_n(Y)$ might be computable in $O(c^n)$ time. Expression (24) for s and the subsequent discussion show that the rational homotopy of a fixed Y can definitely be obtained in $O(c^n)$ time.

We have developed other positive results as well. In the "no news is good news" category, our methods give no lower bound at all on the complexity of the (integral) homotopy groups of spheres, so hope remains for that problem. More affirmatively, our careful consideration of N-encodings suggests that we may have discovered the relative importance of the various inputs. For instance, doubling n has approximately the same effect on the computational difficulty of $\pi_n(X) \otimes Q$ as does squaring the coefficients used in constructing X, since either operation potentially doubles $\ell(I)$. Lastly, Theorems 3.5 and 4.8 greatly expand our repertoire of series known to be rationally related to Hilbert or to Poincaré series.

In summary, computing homotopy groups of arbitrary simply connected spaces is a genuinely computationally difficult problem. We cannot expect that rapid algorithms for it will ever be discovered. Somewhat anticlimactically, we close with Ed Brown's observation to this effect, prophetically written more than 30 years ago [9, p. 1]:

> While the procedures developed for [computing homotopy groups] are finite, they are much too long to be considered practical.

ACKNOWLEDGMENT

I could not have written this chapter if Michael Sipser had not patiently explained to me what $\#\mathcal{P}$ means, and the chapter would not now be in print if Lisa Court and Judy Romvos had not worked so diligently to type it. This work was partially supported by National Science Foundation grant 8509191-DMS.

REFERENCES

1. J. F. Adams and P. J. Hilton, On the chain algebra of a loop space, Comm. Math. Helv. 30, (1955), 305–330.
2. L. Adelman and F. T. Leighton, An $O(n^{1/10.89})$ primality testing algorithm, Math. Comput. 36(153), (1981), 261–266.
3. D. J. Anick, Diophantine equations, Hilbert series, and undecidable spaces, Ann. Math. 122, (1985), 87–112.

4. D. J. Anick, Generic algebras and CW complexes, Algebraic Topology and Algebraic K-Theory (W. Browder, ed.), Annals of Math. Study, 113, Princeton University Press, (1987), 247–321.
5. D. J. Anick, Non-commutative graded algebras and their Hilbert series, J. Algebra 78(1), (1982), 120–140.
6. D. J. Anick and T. H. Gulliksen, Rational dependence among Hilbert and Poincaré series, J. Pure Appl. Algebra 38, (1985), 135–157.
7. H. J. Baues and J. -M. Lemaire, Minimal models in homotopy theory, Math. Ann. 225, (1977), 219–242.
8. D. Bayer and M. Stillman, On the complexity of computing syzygies, preprint.
9. E. H. Brown, Finite computability of Postnikov complexes, Ann. Math. 65, (1957), 1–20.
10. E. Čech, Höherdimensionale Homotopiegruppen, Verhandlungen des Internationalen Mathematikerkongress, Zürich, 1932, Orel Füssli, Zürich (1932), 203.
11. E. B. Curtis, Lower central series of semi-simplicial complexes, Topology 2, (1963), 159–171.
12. E. B. Curtis, Some relations between homotopy and homology, Ann. Math. 82, (1965), 386–413.
13. R. Froberg, T. Gulliksen, and C. Löfwall, Flat families of local Artinian algebras with an infinite number of Poincaré series, Algebra, Algebraic Topology and their Interactions, Lecture Notes in Math., 1183, Springer-Verlag, New York (1986), 170–191.
14. T. H. Gulliksen, Reducing the Poincaré series of local rings to the case of quadratic relations, Algebra, Algebraic Topology and their Interactions, Lecture Notes in Math., 1183, Springer-Verlag, New York (1986), 195–198.
15. J. E. Hopcroft and J. D. Ullman, Introduction to Automata Theory, Languages, and Computation, Addison-Wesley, Reading, Mass. (1979).
16. W. Hurewicz, Beiträge zur Topologie der Deformationen, Nederl. Akad. Wetensch. Proc. Ser. A 38, (1935), 112–119, 521–528; 39, (1936), 117–126, 215–224.

17. C. Jacobsson, Finitely presented graded Lie algebras and homomorphisms of local rings, J. Pure Appl. Algebra 38, (1985), 243–253.

18. D. E. Knuth, Semi-numerical algorithms, The Art of Computer Programming, vol. 2, 2nd ed., Addison-Wesley, Reading, Mass. (1981).

19. G. Levin, Local rings and Golod homomorphisms, J. Algebra 37, (1975), 266–289.

20. S. MacLane, Homology, Die Grundlehren der Mathematische Wissenschaften in Einzeldarstellungen, 114, Springer-Verlag, Berlin (1963).

21. G. L. Miller, Riemann's hypothesis and tests for primality, J. Comput. Syst. Sci. 13(3), (1976), 300–317.

22. J. Milnor and J. C. Moore, On the structure of Hopf algebras, Ann. Math. 81, (1965), 211–264.

23. D. G. Quillen, Rational homotopy theory, Ann. Math. 90, (1969), 205–295.

24. R. Sedgewick, Algorithms, Addison-Wesley, Reading, Mass. (1983).

25. J. -P. Serre, Groupes d'homotopie et classes de groupes abéliens, Ann. Math. 58, (1953), 258–294.

26. C. E. Shannon and J. McCarthy, eds., Automata Studies, Study No. 34, Princeton University Press (1956).

27. D. Sullivan, Infinitesimal computations in topology, Publ. Math. I.H.E.S. 47, (1977), 269–331.

28. L. G. Valient, The complexity of computing the permanent, Theoret. Comput. Sci. 8, (1979), 189–201.

29. K. Weld, Computability of Homotopy Groups of Nilpotent Complexes, Thesis, Graduate School and University Center of City University of New York (1984).

30. G. W. Whitehead, Elements of Homotopy Theory, Springer-Verlag, New York (1978).

2

Geometry of the Hopf Mapping and Pinkall's Tori of Given Conformal Type

THOMAS F. BANCHOFF

Brown University
Providence, Rhode Island

One of the most fertile geometrical examples in mathematics is the Hopf mapping from the 3-sphere S^3 to the 2-sphere S^2. On the occasion of the Conference on Computers in Geometry and Topology at the University of Illinois at Chicago, we considered three aspects of this mapping which are particularly well suited to investigation by computer graphics.

The study of Hamiltonian dynamical systems can be motivated and illustrated by a linear system which leads to the Hopf fibers as circular orbits lying on a constant energy 3-sphere. This aspect has been reported at length elsewhere in the article of Hüseyin Koçak, Fred Bisshopp, David Laidlaw, and the author [1].

Regular polytopes in 4-space can be decomposed into rings of polyhedra which correspond to solid tori which are preimages of cells on the 2-sphere under the Hopf mapping. This aspect has appeared in the author's article in the proceedings of the conference <u>Shaping Space</u> [2].

The third topic considered was an elementary presentation of a remarkable construction by Ulrich Pinkall which determines the conformal structures of tori obtained by lifting a closed curve on S^2 under the Hopf mapping. This short note is an exposition of Pinkall's result which is well-suited to interactive computer graphics investigation.

In [3], Pinkall showed that the inverse image under the Hopf mapping of a simple closed curve on S^2 is a flat torus on S^3 which is conformally equivalent to a parallelogram in the plane with basis vectors

(0,1) and (L/2,A/2), where L is the length of the curve and A is the (oriented) area it encloses on S^2. Stereographic projection then yields embedded tori in R^3 of the same conformal type. This result gives an explicit solution to a problem first solved by Adriano Garsia using methods which were nonconstructive and not suited for the production of examples.

Pinkall's elegant approach utilizes quaternionic multiplication on S^3, arc length parametrization of a curve on S^2, choice of a lifting of the curve which is orthogonal to the Hopf fibers, and calculation of the area by integration of a curvature form on a circle bundle. His article is illustrated by computer-generated line images. In order to investigate these Hopf tori by means of interactive computer graphics, it seems more convenient to use Cartesian coordinates, arbitrary parametrization of the base curve, a specific lift not necessarily orthogonal to the fibers, and an explicit calculation of the area. Fortunately it is possible to carry out Pinkall's argument without introducing any of the technical apparatus of modern differential geometry, thereby producing a proof which, though less elegant, is more elementary, and accessible to students with a knowledge of the classical geometry of curves and surfaces. In this note we give such a direct presentation, together with a set of illustrations indicating the way the stereographic projections of Hopf tori transform under inversions.

We will only need curves on the 2-sphere which are polar coordinate function graphs, $X(\theta) = (\cos(\theta)\sin(\phi(\theta)), \sin(\theta)\sin(\phi(\theta)), \cos(\phi(\theta)))$, where $0 \leq \theta \leq 2\pi$ and where $\phi(\theta)$ is a differentiable function of θ with period 2π and $0 \leq \phi(\theta) \leq \pi$. (Note that these are the coordinates used in astronomy, with ϕ measured down from the North Pole, rather than standard geographical coordinates measuring latitude up and down from the Equator.)

The velocity vector of the curve is given by

$$X'(\theta) = (-\sin(\theta)\sin(\phi(\theta)) + \cos(\theta)\cos(\phi(\theta))\phi'(\theta),$$
$$\cos(\theta)\sin(\phi(\theta)) + \sin(\theta)\cos(\phi(\theta))\phi'(\theta), -\sin(\phi(\theta))\phi'(\theta))$$

Thus the length L(t) is the integral of $\sqrt{X_\theta \cdot X_\theta} = \sqrt{[\sin^2(\phi(\theta)) + (\phi'(\theta))^2]}$ from 0 to t.

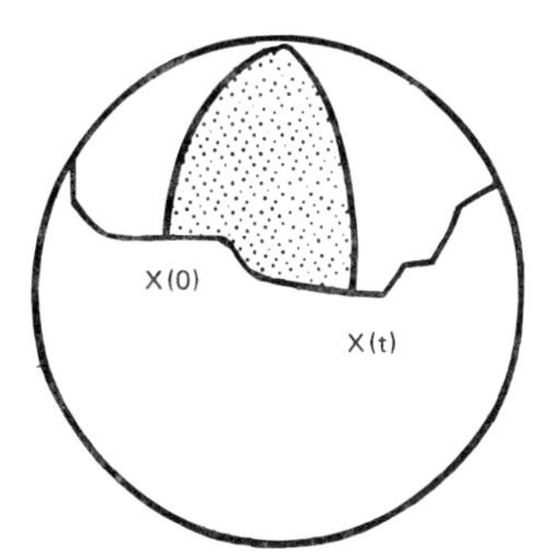

Figure 1

The area A(t) of the wedge from the North Pole down to X(0), along the curve to X(t), and back up to the pole, is the integral of the area element (see Figure 1)

$$|X_\theta \times X_\phi| = \sin(\phi)$$

so

$$A(t) = \int_0^t \int_0^{\phi(\theta)} \sin(\phi)d\phi d\theta = \int_0^t (1 - \cos(\phi(\theta))d\theta$$

To parametrize the unit 3-sphere S^3, we introduce "toral" coordinates $Z(\alpha,\beta,\gamma) = (\cos(\alpha)\sin(\gamma),\sin(\alpha)\sin(\gamma),\cos(\beta)\cos(\gamma),\sin(\beta)\cos(\gamma))$, where $0 \leq \alpha,\ \beta \leq 2\pi$ and $0 \leq \gamma \leq \pi/2$. The Hopf mapping sends a point (x, y,u,v) in R^4 to the point $H(x,y,u,v) = (2xu + 2yv, -2xv + 2yu, u^2 + v^2 - x^2 - y^2)$ in R^3. Thus $H(Z(\alpha,\beta,\gamma)) = (\cos(\alpha - \beta)\sin(2\gamma), \sin(\alpha - \beta)\sin(2\gamma), \cos(2\gamma))$, and the image of S^3 under H is the 2-sphere S^2. The preimage of any point $(\cos(\theta)\sin(\phi),\sin(\theta)\sin(\phi), \cos(\phi))$ is then an entire curve $Z(\theta + \psi,\psi,\phi/2) = (\cos(\theta + \psi)\sin(\phi/2), \sin(\theta + \psi)\sin(\phi/2),\cos(\psi)\cos(\phi/2),\sin(\psi)\cos(\phi/2))$, where $0 \leq \psi \leq 2\pi$. It is easy to show that the preimage of any point is a great circle on S^3. If $X(\theta)$ is a closed curve on S^2 defined by a function $\phi(\theta)$, then these circles fit together to form a torus in S^3, called a Hopf torus. For example, if the curve $X(\theta)$ is given by the function $\psi(\theta) = 2c$, a constant, then the Hopf torus is $(\cos(\theta + \psi)\sin(c),\sin(\theta + \psi)\sin(c),\cos(\psi)\cos(c), \sin(\psi)\cos(c))$, a product torus $Z(\theta + \psi,\psi,2c)$ on S^3.

Some of the motivation for the Hopf mapping becomes clearer if we introduce complex coordinates on S^3, writing a 4-tuple (x,y,u.v) as a pair $[x + iy, u + iv] = [z,w]$ of complex numbers. The Hopf mapping is then defined by sending the pair $[z,w]$ to the ratio w/z if $z \neq 0$ and ∞ if $z = 0$, thus giving a point on C with ∞ adjoined. Inverse stereographic projection (plus a reflection) then takes w/z to $[2z\bar{w}, w\bar{w} - z\bar{z}]$ and takes ∞ to $[0,1]$. In polar coordinates, $Z(\alpha,\beta,\gamma) = [\sin(\gamma)e^{i\alpha}, \cos(\gamma)e^{i\beta}]$, and $H[\sin(\gamma)e^{i\alpha}, \cos(\gamma)e^{i\beta}] = [\sin(2\gamma)e^{i(\alpha-\beta)}, \cos(2\gamma)]$.

The claim is that any Hopf torus is flat, with Gaussian curvature identically zero. The easiest way to establish this is to show that this surface is isometric to a specific region in the plane. We first compute the metric coefficients of the surface $Y(\theta,\psi) = Z(\theta + \psi, \psi, \phi(\theta)/2)$:

$$
\begin{aligned}
Y_\theta = (&-\sin(\theta + \psi)\sin(\phi(\theta)/2) + \cos(\theta + \psi)\cos(\phi(\theta)/2)\phi'(\theta)/2,\\
&\cos(\theta + \psi)\sin(\phi(\theta)/2) + \sin(\theta + \psi)\cos(\phi(\theta)/2)\phi'(\theta)/2,\\
&-\cos(\psi)\sin(\phi(\theta)/2)\phi'(\theta)/2,\\
&-\sin(\psi)\sin(\phi(\theta)/2)\phi'(\theta)/2),\\
Y_\psi = (&-\sin(\theta + \psi)\sin(\phi(\theta)/2), \cos(\theta + \psi)\sin(\phi(\theta)/2),\\
&-\sin(\psi)\cos(\phi(\theta)/2), \cos(\psi)\cos(\phi(\theta)/2))
\end{aligned}
$$

Then $g_{11} = Y_\theta \cdot Y_\theta = \sin^2(\phi(\theta)/2) + (\phi'(\theta)/2)^2$, $g_{12} = Y_\theta \cdot Y_\psi = \sin^2(\phi(\theta)/2)$, and $g_{22} = Y_\psi \cdot Y_\psi = 1$.

We now define a planar region by $W(\theta,\psi) = (L(\theta)/2, A(\theta)/2 + \psi)$. Then $W_\theta = (L'(\theta)/2, A'(\theta)/2)$, $W_\psi = (0,1)$. The metric coefficients are then $G_{11} = W_\theta \cdot W_\theta = L'(\theta)^2/4 + A'(\theta)^2/4$, $G_{12} = W_\theta \cdot W_\psi = A'(\theta)/2$, and $G_{22} = W_\psi \cdot W_\psi = 1$. First note $G_{22} = g_{22}$. Since $A'(\theta)/2 = [1 - \cos(\phi(\theta))]/2 = \sin^2(\phi(\theta)/2)$, we also have $G_{12} = g_{12}$. Finally, $L'(\theta)^2/4 + A'(\theta)^2/4 = \sin^2(\phi(\theta))/4 + \phi'(\theta)^2/4 + \sin^4(\phi(\theta)/2) = \sin^2(\phi(\theta)/2)\cos^2(\phi(\theta)/2) + \phi'(\theta)^2/4 + \sin^2(\phi(\theta)/2)\sin^2(\phi(\theta)/2)$, so $G_{11} = g_{11}$. Therefore the Hopf torus has precisely the same metric coefficients as a region in the plane bounded by two vertical segments of length 2π and two other curves which are parallel translates of one another. This identification space is isometric to the parallelogram with vertices at (0,0), (L/2,A/2), $(L/2 + 2\pi, A/2)$, and $(0,2\pi)$ (see Figure 2).

Pinkall shows that it is possible to choose curves on the sphere which yield all possible parallelograms with third vertex in a given fun-

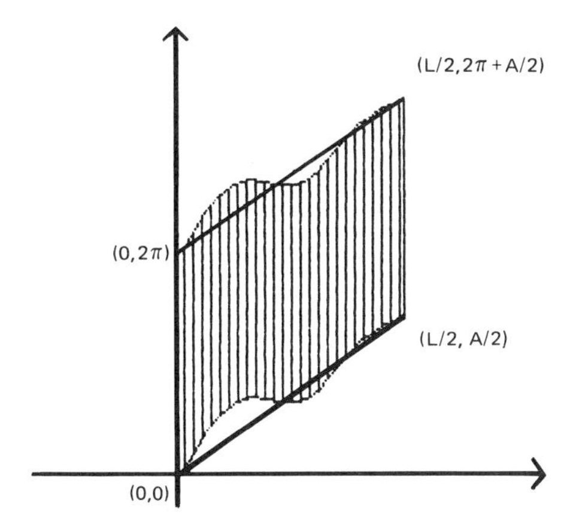

Figure 2

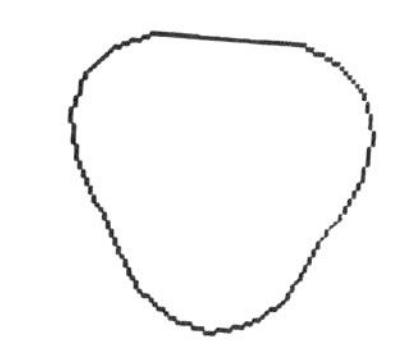

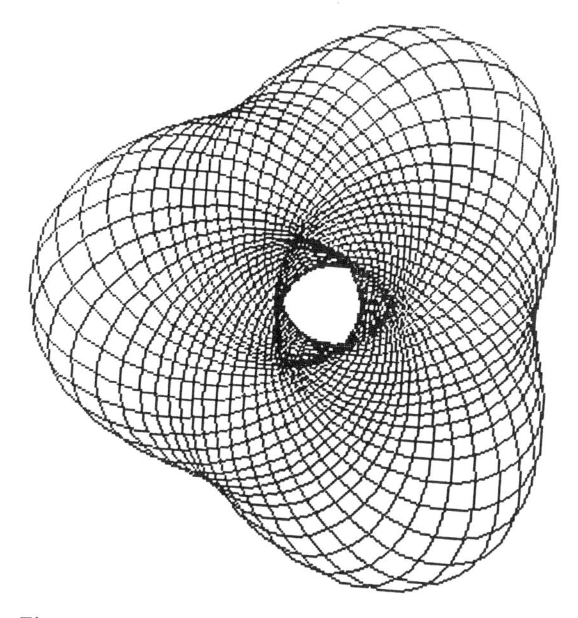

Figure 3

damental region, so this completely solves the problem of finding tori on the 3-sphere of given conformal types. Stereographic projection gives embedded tori in ordinary 3-space with the same conformal types (see Figure 3).

Added in Proof: Joel Weiner has been able to use the techniques of this paper to extend his results on Flat Tori in S^3 and Their Gauss Maps.

REFERENCES

1. T. Banchoff, (with H. Koçak, F. Bisshopp, and D. Laidlaw), Topology and Mechanics with Computer Graphics: Linear Hamiltonian Systems in Four Dimensions, Advances in Applied Mathematics, (1986), 282–308.
2. T. Banchoff, Torus Decompositions of Regular Polytopes in 4-Space, Shaping Space, (1988), Birkhäuser, Boston, 221–230.
3. U. Pinkall, Hopf Tori in S^3, Inventiones Mathematicae, 81, (1985), 379–386.

3

Environments
An Algebraic Computing Technique

MAX BENSON

University of Minnesota
Duluth, Minnesota

An environment is a representation of an algebraic structure within a program. This chapter describes a programming technique for algebraic computing which is based on environments. Support routines are being developed by the author for use in defining and programming with environments from within the C language. The final section gives an example of a working program written using this technique which computes the Todd polynomials $T_n(y;c_1,...,c_n)$.

1. INTRODUCTION

Algebraic computations have long been performed by computers. Early efforts involved writing a program to make a specific computation using a conventional programming language such as FORTRAN. Because the algebraic objects that mathematicians are interested in are not among the primitive data types that are provided by most programming languages, the overhead of maintaining the data structures used for internal storage of algebraic objects and implementing the primitives for these objects is significant. Parsing, output, and garbage collection all present challenges.

General-purpose systems such as MACSYMA, REDUCE, SMP, and MAPLE provide an alternative approach. They can evaluate and simplify expressions entered by the user. They also offer programmability in a high-level language. Most of the implementation details are hidden.

The environments provided by these systems have satisfied many of the computer algebra needs of mathematicians because of their ease of use.

These systems do have shortcomings. At least one has been criticized for its errors [6]. And in most cases the execution time of a native program designed for the calculation would be shorter. Nevertheless, the complexity of writing such a program means that using one of these systems is more likely to give a correct and timely answer to the problem.

That is, if they are available to the mathematician. They are large programs and may be restricted by their size to a large computer. Some only run under certain operating systems. Licensing fees may keep them running on only a few, if any, of the institution's computers. Either way, the mathematician may be forced to use an unfamiliar computer that costs real dollars to use.

More serious is the number of the problems that these systems just can't solve very well. Sometimes the lack of control over the internal data structures means that the system can't perform some types of symbol manipulation well. An example illustrating this is given in the last section.

For algebraic geometers and topologists the major problem is that they do symbol manipulation instead of abstract algebra. They treat algebraic objects as symbolic quantities and not as elements of a ring or module. A new system, SCRATCHPAD II, may address these needs, but it is not a product yet [4]. Some needs are being met by special-purpose algebra programs such as Macaulay [2].

This chapter gives an alternative to using the existing symbolic manipulation systems. It describes an object-oriented programming technique that can be used with the widely available C language to create algebraic data types and compute with them. Unlike the systems mentioned above, it is only a programming technique aided by libraries of support routines, but it has been effective in solving problems and may be a useful foundation for new computer algebra systems.

2. ENVIRONMENTS

Our goal is to be able to perform operations on elements of algebraic structures such as rings and modules instead of just storing expressions and manipulating the symbols within them. Following the object-oriented style of programming, we represent an algebraic structure by a template for how to store one of its elements in computer memory along with the code for the primitive operations that can be performed on elements within it. Such a representation of an algebraic structure is called an environment.

In the following subsections we describe seven different constructions that can be used to produce environments representing many different algebraic structures. The first three constructions produce building blocks which the other four can put together to represent complicated algebraic structures. All of the data structures they use are familiar.

2.1 Exact Integer Environment

Elements of the ring of exact integers, Z, can be represented base 2^m, where m is half the word size of the computer. The data structure is depicted below. See [5] for some of the implementation details.

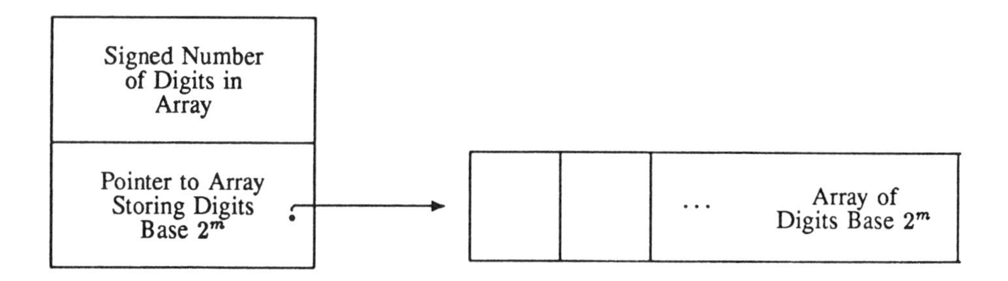

The exact integer environment consists of this data structure plus the code for the primitive operations. These operations include arithmetic primitives such as addition, negation, subtraction, multiplication, division, and exponentiation as well as others like conversion between machine integers and exact integers, comparison, and output.

2.2 Monic Monomials Environment

Monomials in the variables $x_1, x_2, \ldots, x_n$ with a coefficient of 1 can be represented by an array of length n specifying their exponents

Exponent of x_1	Exponent of x_2	...	Exponent of x_n

together with the code for functions which perform such operations as multiplication, division, exponentiation, comparison, and output. Another important primitive is a constructor function which takes a string representing one of the variables x_i and constructs a data object storing it.

By varying the comparison function the ordering on the monomials may be changed. Reversing its sign makes a descending order an ascending one. For monomials in several variables, weights may be placed on the variables and either lexicographic order or reverse lexicographic order can be used.

2.3 Signed Monic Wedge Form Environment

The set of wedge forms with coefficients ±1 can be represented by a record of the following form:

Sign of Form
Number of Symbols in this Wedge Form
Bit Mask indicating which Symbols are present

along with implementations of primitive operations like negation, multiplication, division, construction, and output. The information stored in the record is redundant, but it tends to simplify some of the routines.

The next construction makes possible the representation of important algebraic structures such as polynomial and power series rings and of exterior algebras.

2.4 Adjunction Environments

Suppose R is a commutative ring with unity and G is an ordered set with a product $G \times G \to G \cup \{0\}$ that has a unit element. The set R[G] of formal sums $\sum r_i g_i$, r_i R, $g_i \in G$ can be endowed with a natural ring structure.

If both R and G are represented by environments, then we can construct an environment to represent R[G]. An element of R[G] can be represented by a sorted linked list of records. Each node of the list represents one term in the formal sum, with an empty list meaning 0. The nodes are be sorted according to the ordering of G.

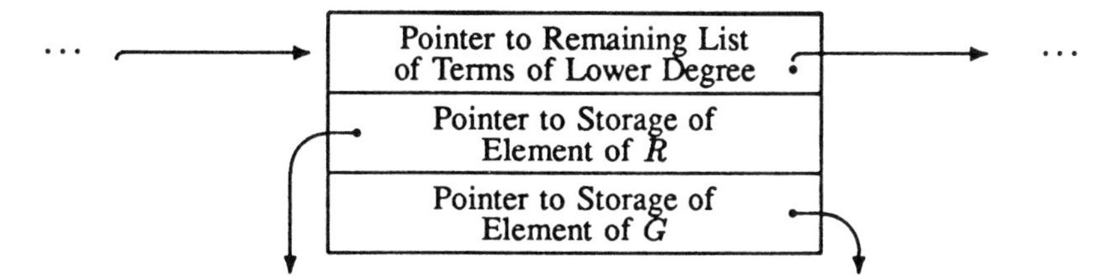

The primitive operations for adjunctions call on the primitives of R and G. But the algorithms are independent of specific choices of R and G.

2.5 Ring of Fractions Environments

Given a commutative ring R and a multiplicative set $S \subset R$, the ring $S^{-1}R$ consists of formal fractions with numerators in R and denominators restricted to S. A special case is when R is an integral domain and $S = R^*$. Then $S^{-1}R$ is the fraction field of R.

In order to construct an environment for $S^{-1}R$, we need a representation of R and two special primitives. The first is a boolean predicate in the environment of $S^{-1}R$ which determines whether an element of R is in S. The second is a primitive in the environment of R which takes two elements of R and reduces the fraction formed by them to lowest terms. This is used to keep fractions stored in a reduced form.

A data structure consisting of a record with pointer fields for the storage of the numerator and denominator can be used to store an element of $S^{-1}R$.

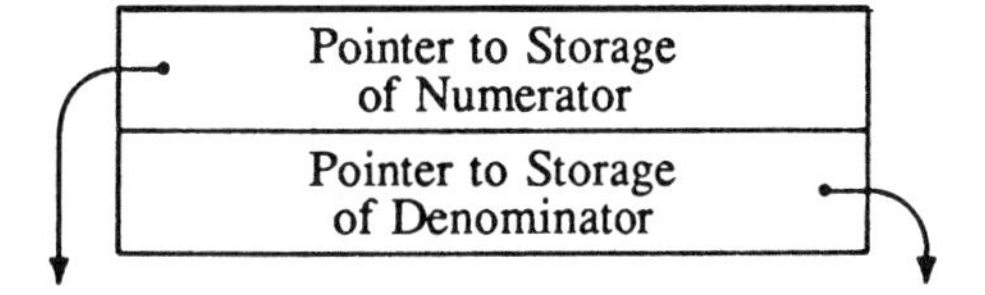

2.6 Quotient Ring Environments

Suppose that R is a commutative ring and $I \subset R$ is an ideal. Given an environment representing R, there are two problems to solve before creating an environment for R/I.

Each element of R/I is an equivalence class of elements of R. We need to choose a unique representative element in R for each class in R/I and be able to reduce any element of R to its representative.

In the important cases, the reduction algorithm does exist. If R has a division algorithm and I is a principal ideal, then this reduction process only involves taking the remainder after dividing by the generator of I. Even if I is not principal, in polynomial and power series rings it is possible to first prepare the ideal by finding a new set of generators. The remainder obtained by performing a generalized division procedure with respect to the new set of generators is the representative that should be used [1].

The second problem is the existence of an inversion algorithm. For the rings Z/mZ, Euclid's algorithm can be used. For quotients of polynomial rings, power series rings, and exterior algebras, inversion is more difficult.

This construction differs from the others in that no additional storage is needed for the element. The environment is needed for bookkeeping in order to ensure that reductions are made at the right time in the computation.

Together, these six constructions make it easy to construct environments for many familiar algebraic structures. A representation for the ring $Q(t)[[x,y]]/(x^2 - y^3)$ is depicted below.

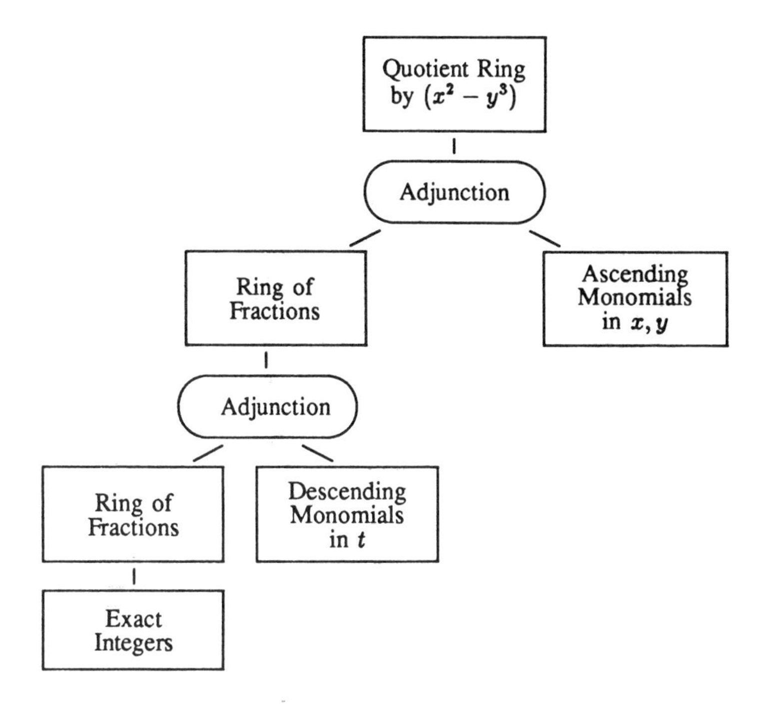

2.7 Sparse Tuple Family of Environments

Sparse tuples are linked lists of records which store an integer index and a pointer to an item. Items which are zero are omitted. Their presence is indicated by the missing index. There is a special header node which has one field pointing to the start of the list and a second field that stores a position in the list.

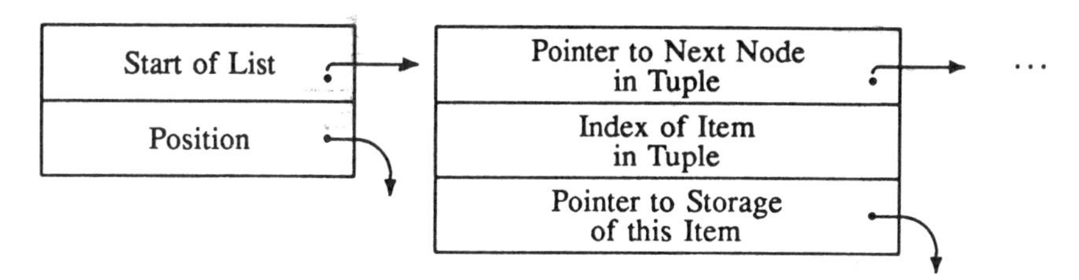

By varying the implementations of the primitives, many different environments can be defined.

Some sparse tuple environments are for storing lists. The position pointer is used explicitly in list operations such as deletion and retrieval. An indexed list environment can be constructed which allows insertion at any index that is specified. A variation is the sorted list environment, which ignores the index and allows only sorted insertion. List environments are quite useful. Ideals are represented by lists of generators. A system of linear equations can be stored as a list sorted on the index of the first nonzero coefficient.

Other environments are more algebraic. There are no list primitives, and the position pointer is not used explicitly. However, the implementations of the operations attempt to leave the pointer in the right place so that in many cases the next call on the operation can continue processing without first locating the correct position in the list.

The simplest algebraic objects that can be represented are vectors and linear equations. Matrices can be viewed as tuples of vectors as long as the proper code for multiplication and inversion is added. Carrying this further, we have tensors.

Another use is to represent finite-dimensional algebras such as Lie algebras. The multiplication routine multiplies componentwise, checking a matrix of tuples which stores the products of the basis elements.

3. IMPLEMENTATION

The C language has been chosen for implementation because of its wide availability, portability, and efficiency. The programming technique proposed in this chapter is very language dependent, and these last two sections discuss environments from the context of the C language.

Elements of an environment are stored as pointers to the data structures described in the last section. All pointers are made equivalent by placing them in a <u>union</u>. The type declaration is shown below.

```
typedef union element {
    int integer ;
    struct exact * exact ;
    struct fraction * fraction ;
    struct adjoin * adjoin ;
    short * monomial ;
    struct wedge * wedge ;
    struct tuple * tuple ;
} ELEMENT ;
```

The record declarations for the actual data structures are not shown.

An environment is not just an abstraction. It is stored explicitly in a tree structure which mimics the storage of the actual elements, except that some nodes, like those for reduction, appear in the environment tree but not in the storage of the element. Each node in the environment tree is a very large record which stores tree pointers, environment-specific data such as the number and names of variables, and pointers to C functions implementing the primitives at that level. Here is a shortened version of this record.

```
typedef struct environment {
    struct environment * co_environment ;
    struct environment * term_environment ;

    char ** Names ;
    short * WeightVector ;
    short NumberNames ;
    short * Groups ;
    short NumberGroups ;

    int (*_Initialize)() ;
    int (*_BuildNumber)() ;
    int (*_BuildVariable)() ;
    int (*_BuildInteger)() ;
    int (*_IntegerRestrict)() ;
    int (*_ExtendScalar)() ;
    int (*_ScalarRestrict)() ;
    int (*_ExtendtoHull)() ;
    int (*_Restrict)() ;
    int (*_Copy)() ;
    int (*_Destroy)() ;
    int (*_Add)() ;
    int (*_Subtract)() ;
    int (*_Multiply)() ;
    int (*_IntMultiply)() ;
    int (*_IntQuotient)() ;
    int (*_ScalMultiply)() ;
    int (*_ScalQuotient)() ;
    int (*_Quotient)() ;
    int (*_Raise)() ;
    int (*_Negate)() ;
    int (*_Put)() ;
    int (*_Compare)() ;
    int (*_EqualsZero)() ;
    int (*_LeadTerm)() ;
    int (*_LeadCoef)() ;
    int (*_Exponent)() ;
} ENVIRONMENT ;
```

The real version has over 80 primitive operations. Some of the primitives that have been omitted perform elimination, division, differentiation, and list operations.

In order to call one of the primitives, say _Negate from the environment E, the awkward syntax (E->_Negate)(a,b,E) would have to be used. For this reason there is a macro defined for each primitive to improve readability. negate(a,b,E) is the macro version of the _Negate operation.

Generally, the primitives take arguments, perform an operation, and store the results back into a parameter. Depending on the particular operation, return values are either meaningless or indicate the status of the operation. For example, the primitive _Negate places the negation of a in the element pointed to by b. The element formerly stored by b is gargabe-collected. The return value is ignored.

Of course, the pointers to the primitives refer to actual routines. Consider the environment representing Z[x,y]. At the top-level node, _Negate points to the routine aneg, which negates adjunctions. Here is a possible implementation of this C function.

```
aneg( a, b, E )
ELEMENT a, *b ;
ENVIRONMENT * E ;
{
    ELEMENT ptr ;

    copy( a, b, E ) ;
    for( ptr = *b ; ptr.adjoin != (ADJOIN *)NULL ; ptr.adjoin = ptr.adjoin->rest )
        negate( ptr.adjoin->coefficient, &ptr.adjoin->coefficient, E->co_environment ) ;
}
```

This routine first invokes the primitive _Copy, which garbage-collects b and assigns it a pointer to new memory containing a copy of a. Then it traverses the newly created linked list. At each node along the traversal, it calls xneg, the negation primitive of the exact integers, to change the sign of coefficient.

In the node representing the monomials in x,y, _Negate points to the routine noexist3. Negation of monomials is not permitted, and the purpose of this routine is to print a diagnostic error message.

Before storing elements of some environment the programmer has to build the tree structure for the environment in memory. This is not difficult because special construction routines have been provided. We will describe the ones that appear in the next section. All of them return a pointer to the tree structure they create. makeZ( ) creates an exact integer environment. makeadjoin(coef, term) constructs an adjunction environment given pointers to the environments of the coefficients and terms. makefraction(R, member_S), builds an environment for $S^{-1}R$, given a pointer to the environment R and a pointer to a function, member_S, which checks whether an element is in S. Finally makemonomial(names, groups, weights, number_names, number_groups, attributes) creates a monomial environment. Weights can be specified by an array of integers. A bit vector indicates such attributes as ASCENDING and REVERSE_LEX. Variables can be placed in groups to make elimination orderings possible.

Once the environments are created, the programmer may use variables of type ELEMENT to store algebraic objects. Algorithms can be encoded using the macros for the primitives.

Care must be taken so that objects are initialized properly, because many routines, such as _Negate, garbage-collect the space that output parameters point at before assigning the result to the parameter. The _Initialize primitive can be used to simply set an object to NULL. _BuildNumber and _BuildVariable also perform initialization. They take a character representation of a number or a variable, search the environment tree for the node where it appears, and then construct an internal representation for the object in that subenvironment.

The programmer also has to make sure that pointers do not get corrupted by the garbage collection process. This is avoided by not using more than one pointer to refer to the same object. A primitive, _Destroy, gives the programmer the ability to specify that an object should be garbage-collected if no longer needed.

The _Put operation can be used to print an object on a stream. Additional parameters control the printing of parentheses, plus signs, and unit coefficients. On the other hand, parsing routines are not part of the primitive operations because input formats and error-handling behavior are so variable. It is not hard to construct a parser using _BuildNumber, _BuildVariable, and the algebraic primitives.

A common problem is finding that the objects are stored in the wrong environment. One solution is to use one of the many conversion primitives. Machine integers can be converted to exact integers by the _BuildInteger operation. Similarly, a representation of an element in a subenvironment may be changed to a representation in an enclosing environment using _ExtendScalar. For instance, this routine could extend the representation of "2" as an exact integer to a representation as a polynomial. The primitives _IntegerRestrict and _ScalarRestrict make conversion possible in the other direction. More generally, there is an operation of _Retract which replaces a representation of an element by an equivalent one in the smallest environment possible. There is also a fairly complicated primitive _ExtendToHull which extends an object to the smallest enclosing environment which has certain specified properties.

It is not always necessary to make the conversion. The primitives _IntMultiply, _IntQuotient, _ScalMultiply, and _ScalQuotient give one the ability to multiply or divide some element by machine integers or by elements in subenvironments.

4. EXAMPLE

The implementation described in the last section is being developed using SUN-2 and FORTUNE 32:16 computers. Both run UNIX*, but the versions of the operating system and the C compilers on the two machines are quite different. This has been useful for finding portability problems.

Although the implementation process is not complete, most of what is described in this chapter is already working. In this section we give an example of a working program which uses many of these primitives. This program was written to aid two colleagues, Anatoly Libgober and John Wood of the University of Illinois at Chicago, who have been interested in calculating the Todd polynomials $T_n(y;c_1,\ldots c_n)$.

They proposed an algorithm to compute the Todd polynomials based on a classic method used to show that a symmetric polynomial can be expressed as a polynomial in the elementary symmetric functions [7]. Here is description of how the process works.

$T_n(y;c_1,\ldots,c_n)$ is a weighted homogeneous polynomial of degree n in $c_1,\ldots c_n$ where c_i is assigned the weight i. The T_n form a multiplicative sequence, and the coefficients of the characteristic power series $Q(y;x) = 1 + q_1x + q_2x^2 + \cdots$ can be calculated. Each coefficient q_i is a polynomial in y of degree i. $T_0(y) = 1$ and the general T_n are given by the recurrence formula [3]

*UNIX is a registered trademark of AT&T.

$$T_n(y;\gamma + c_1,\gamma c_1 + c_2,\ldots,\gamma c_{n-1} + c_n) - T_n(y;c_1,\ldots,c_n) = q_n\gamma^n + \cdots + q_1\gamma T_{n-1}(y;c_1,\ldots,c_{n-1})$$

The right side of this formula can be calculated, but both terms on the left-hand side are unknown. Order the right-hand side lexicographically. The lead term will be of the form $a\gamma^{e_0}c_1^{e_1}\cdots c_{n-1}^{e_{n-1}}$. It must come from a term of the form $a(\gamma + c_1)^{e_0-e_1-\cdots-e_{n-1}}(\gamma c_1 + c_2)^{e_1}\cdots(\gamma c_{n-1} + c_n)^{e_{n-1}}$ appearing in $T_n(y;\gamma + c_1,\gamma c_1 + c_2,\ldots,\gamma c_{n-1} + c_n)$. Subtract this term off from both sides of the equation. Now we can start over and apply the same reasoning. This process of subtracting off appropriate terms continues until the right-hand side does not contain anything involving γ. Then the left-hand side is just $-T_n(y;c_1,\ldots,c_{n-1})$.

Wood implemented this algorithm with a REDUCE program on the campus mainframe. Unfortunately, the version of REDUCE available did not offer a convenient method of storing polynomials in weighted lexicographic order. The resulting program had to search for the lead term in a very awkward manner, and it required too much computer time to complete the calculations.

The author wrote the following C program to compute T_n for $2 \leq n < 9$.

```
#include <stdio.h>
#include "../h/element.h"
#include "../h/environment.h"
#include "../h/decls.h"

extern ELEMENT parse() ;

ENVIRONMENT * exact, * fraction, * zvariables, * cvariables ;
ENVIRONMENT * zpolynomials, * cpolynomials ;

char * znames[] = { "z" } ;
char * cnames[] = { "g", "c1", "c2", "c3", "c4", "c5", "c6", "c7", "c8" } ;

short weights[] = { 1, 1, 2, 3, 4, 5, 6, 7, 8 } ;

ELEMENT var[8] ;

main() {
    ELEMENT q[9], t[9] ;
    ELEMENT product ;
    short i, j ;

    exact = makeZ() ;
    fraction = makefraction( exact, not_zero ) ;
```

```
    zvariables = makemonomial( znames, (short *)NULL, (short *)NULL, 1, 1, 0 );
    zpolynomials = makeadjoin( fraction, zvariables ) ;
    cvariables = makemonomial( cnames, (short *)NULL, weights, 9, 1, 0 );
    cpolynomials = makeadjoin( zpolynomials, cvariables ) ;

    q[0] = parse( "1", cpolynomials ) ;
    q[1] = parse( "(1 - (1/2)z)g", cpolynomials ) ;
    q[2] = parse( "(1/12)z^2 g^2", cpolynomials ) ;
    q[3] = parse( "0", cpolynomials ) ;
    q[4] = parse( "-(1/720)z^4 g^4", cpolynomials ) ;
    q[5] = parse( "0", cpolynomials ) ;
    q[6] = parse( "(1/30240)z^6 g^6", cpolynomials ) ;
    q[7] = parse( "0", cpolynomials ) ;
    q[8] = parse( "-(1/1209600)z^8 g^8", cpolynomials ) ;

    var[0] = parse( "g + c1", cpolynomials ) ;
    var[1] = parse( "g c1 + c2", cpolynomials ) ;
    var[2] = parse( "g c2 + c3", cpolynomials ) ;
    var[3] = parse( "g c3 + c4", cpolynomials ) ;
    var[4] = parse( "g c4 + c5", cpolynomials ) ;
    var[5] = parse( "g c5 + c6", cpolynomials ) ;
    var[6] = parse( "g c6 + c7", cpolynomials ) ;
    var[7] = parse( "g c7 + c8", cpolynomials ) ;

    t[0] = parse( "1", cpolynomials ) ;
    t[1] = parse( "(1 - (1/2)z) c1", cpolynomials ) ;

    initialize( &product, cpolynomials ) ;
    for ( i = 2 ; i < 9 ; i ++ ) {
        initialize( &t[i], cpolynomials ) ;
        for ( j = 0 ; j < i ; j ++ ) {
            multiply( q[i-j], t[j], &product, cpolynomials ) ;
            add( t[i], product, &t[i], cpolynomials ) ;
        }
        findTodd( &t[i] ) ;
        negate( t[i], &t[i], cpolynomials ) ;
        printf( "t[%d] = ", i) ;
        put( stdout, t[i], 1, 0, 0, 0, cpolynomials ) ;
        putchar( '\n' ) ;
        putchar( '\n' ) ;
    }
}

findTodd( f )
ELEMENT * f ;
{
    ELEMENT product, power, term, coef ;
    short exp, i, sum ;

    initialize( &product, cpolynomials ) ;
    initialize( &power, cpolynomials ) ;
    while( !equals_zero( *f, cpolynomials ) ) {
        lead_term( &term, *f, cpolynomials ) ;
        exponent( &sum, term, 0, cvariables ) ;
        if ( sum == 0 )
            break ; ;
        for ( i = 1 ; i < 8 ; i ++ ) {
            exponent( &exp, term, i, cvariables ) ;
            sum -= exp ;
        }
        raise( var[0], sum, &product, cpolynomials ) ;
        for ( i = 1 ; i < 8 ; i ++ ) {
            exponent( &exp, term, i, cvariables ) ;
            if ( exp != 0 ) {
                raise( var[i], exp, &power, cpolynomials ) ;
                multiply( power, product, &product, cpolynomials ) ;
            }

        }
        lead_coef( &coef, *f, cpolynomials ) ;
        scalar_multiply( coef, zpolynomials, &product, cpolynomials ) ;
        subtract( *f, product, f, cpolynomials ) ;
    }
    destroy( &product, cpolynomials ) ;
    destroy( &power, cpolynomials ) ;
}
```

This program calls on a routine parse which takes a character string and an environment as arguments and returns an ELEMENT storing the element it represents. This routine is not shown, but it has been written using the UNIX compiler development tools yacc and lex.

This program was compiled on a SUN 2/120 microcomputer with 6 megabytes of memory and linked with the support libraries. It ran successfully in the elapsed time of 2 minutes and 20 seconds.

ACKNOWLEDGMENTS

Much of the work on this project was completed while the author was visiting the Department of Mathematics, Statistics, and Computer Science at the University of Illinois in Chicago during the spring of 1986. The author appreciates the generous support. He also wishes to thank Stephen Yau and his family for their hospitality during his stay.

REFERENCES

1. David Bayer, The Division Algorithm and the Hilbert Scheme, Ph.D. Thesis, Harvard University (June 1982).
2. David Bayer and Michael Stillman, The design of Macaulay: A system for computing in algebraic geometry and commutative algebra (January 1986).
3. F. Hirzebruch, Topological Methods in Algebraic Geometry, Springer-Verlag, New York (1966), 13–15.
4. R. D. Jenks, A primer: 11 keys to new scratchpad, EUROSAM 84, Springer-Verlag, New York (1984), 123–147.
5. David Kingsley, Exact Integer Multiplication Algorithms, Technical Report 86-8, Department of Computer Science, University of Minnesota at Duluth (May 1986).
6. Michael Monagan, Gaston Gonnet, and Bruce Char, Symbolic mathematical computation, Commun. ACM, 29(7), (1986), 680–682.
7. B. L. van der Waerden, Algebra, I, Springer-Verlag, Berlin (1971), 100–101.

4

Calculation of Large Ext Modules

ROBERT R. BRUNER

Wayne State University
Detroit, Michigan

1. INTRODUCTION

Let A be a graded algebra of finite type over a finite prime field k, and let M be an A-module of finite type which is bounded below. Our goal is to compute

$$H^{st}(A) = \mathrm{Ext}_A^{st}(k,k) \qquad \text{and} \qquad H^{st}(M) = \mathrm{Ext}_A^{st}(M,k)$$

for small s and t by directly constructing free resolutions

$$0 \leftarrow M \leftarrow D_0 \leftarrow D_1 \leftarrow D_2 \leftarrow \cdots$$

and

$$0 \leftarrow k \leftarrow C_0 \leftarrow C_1 \leftarrow C_2 \leftarrow \cdots$$

We then compute the action of H(A) on H(M) by constructing a chain map $\tilde{x} : D \to C$ lifting a cocycle $x : D \to k$ representing each H(A) indecomposable $x \in H(M)$. Finally, we compute Massey products $\langle a,b,c \rangle \in H(M)$ by computing a chain null homotopy of bc, for $a,b \in H(A)$ and $c \in H(M)$ satisfying $ab = 0$ and $bc = 0$.

In the programming, our main goal has been economy of space, since memory has been the limiting factor in our experience. A second goal has been to make it as easy as possible to use different algebras and modules. The latter goal has been achieved by using a very small number of routines to define the algebra and the module, so that a new algebra or module can be used by simply redefining these few routines.

We first describe the algorithms for doing the calculations outlined above, then describe our implementations.

2. COHOMOLOGY

2.1 The Connected Case

Assume first that A is connected ($A_0 = k$). Then modules have minimal free resolutions, and an A basis for the minimal resolution of M is dual to a k basis for H(M). To find the minimal resolution, apply the following algorithm, stopping when you want to or have to.

```
Ext := 0;
For t = conn(M) to ∞ begin
    Oldker := Basis for M_t;
    For s = 0 to Maxfilt(t) begin
        Image := ∅;
        Newker := ∅;
        For g ∈ Ext^s
            For op ∈ Basis of A_{t-deg(g)} begin
                x := op * g;
                dx := Act(op,diff(g));
                Reduce (x,dx) against Image;
                If dx = 0 then append x to Newker
                          else insert (x,dx) into Image;
            end {op and g};
        For cyc ∈ Oldker begin
            Reduce cyc against Image;
            If cyc ≠ 0 then begin
                add a generator g to Ext^s;
                diff(g) := cyc;
                deg(g) := t;
                insert (g,cyc) into Image;
        end {cyc};
        Oldker := Newker;
    end {s};
end {t}.
```

Several parts of this require some elaboration or definition.

1. Ext^s is a list of generators in filtration s, for each $s \geq 0$. Tor would probably be more appropriate, but the perfect duality between Tor and Ext here allows us to name it after the actual object of interest to us.
2. $\mathrm{conn}(M) = \min\{t \mid M_t \neq 0\}$, the connectivity of M.

3. diff(g) and deg(g) are the differential and degree of the generator g.
4. Oldker and Newker are lists of elements which span the kernels.
5. Image is an ordered list of pairs (x,dx) such that the dx's form a basis for the image of D^{st} in $D^{s-1,t}$.
6. Maxfilt(t) must be a nondecreasing function of t for the results of the algorithm to have any significance. Within this requirement, Maxfilt may be adapted to the needs or knowledge of the user. Occasionally, we will want only a presentation

$$D_1 \to D_0 \to M \to 0$$

in which case Maxfilt(t) = 1 is appropriate. If a complete resolution is required and nothing is known about M or H(M), Maxfilt(t) = t − conn(M) is appropriate. If we know that $H^s(M) = 0$ for $s > U(t)$ then Maxfilt(t) = U(t) is appropriate. Letting s go any higher than this will merely verify the "vanishing line" U(t). In our original application, A = the mod 2 Steenrod algebra and $M = Z_2$, an interesting variant of this occurs. We have $H^{st}(A) = 0$ for $s > t/3$, approximately, except for $H^{tt}(A)$, which has 1 generator whose differential is Sq^1 of the generator in $H^{t-1,t-1}(A)$. Rather than waste two-thirds of our effort as we would using Maxfilt(t) = t, we "prime" the algorithm with the generators in H^{tt} and let Maxfilt(t) = t/3 (approximately). A similar procedure can be applied whenever D^{st} can be determined from prior values of D. Note that it is not sufficient to know H^{st} in terms of prior H; we also need to know the differential $d : D^{st} \to D^{s-1,t}$ since these differentials may well play a role in higher internal degrees.
7. Act(op,diff(g)) applies op to diff(g). If s = 0, diff(g) $\in$ M, so this depends on M. If s > 0 then diff(g) $\in$ D, so this depends only on the algebra A (since D is free).
8. "Reduce (x,dx) against Image" runs through Image in order, replacing (x,dx) by (x − a,dx − da) for each pair (a,da) $\in$

Image such that the leading term of da occurs in dx, until the leading term of dx is distinct from the leading terms of all the entries in Image, or until dx becomes 0.

"Reduce cyc against Image" runs through Image in order, replacing cyc by cyc − da for each pair (a,da) ∈ Image such that the leading term of da occurs in cyc, until the leading term of cyc is distinct from the leading terms of all the entries in Image, or until cyc becomes 0.

9. Append simply puts the new element at the end of Newker, because we have no need to order the bases for the kernels.
10. Insert inserts (x,dx) into Image so that the dx's are in order, where we order elements by their leading terms. Conceptually, this amounts to keeping the matrix representation of the differential in upper triangular (or row echelon) form with respect to the appropriate bases.
11. During the second phase ("For cyc ∈ Oldker...") we compute Ext^{st} by adding generators as necessary to make the cokernel

    $$\text{Image} \to \text{Oldker} \to \text{Coker} \to 0$$

    equal 0. We do this by making each element of Oldker "orthogonal" to Image using Reduce. Here, orthogonal means that the leading terms of the basis for the Image do not occur in the element of Oldker.
12. The parts of the algorithm which do not specify an order in which to process elements do not depend on the order for their correctness. However, careful attention to the order can have significant effects on the speed with which Reduce operates. Basically, we want to produce (op,gen) pairs in an order that will cause the leading terms of op * diff(gen) to appear in reverse order.

2.2 The Nonconnected Case

Here we do not have resolutions that are minimal in the sense that the differentials become 0 when $- \otimes_A k$ or $\mathrm{Hom}_A(-,k)$ is applied. A simple example is provided by $A = Z_2[Z_3]$ concentrated in degree 0. We have

$H^s(A) = 0$ for $s > 0$, but there are no free resolutions of Z_2 over $Z_2[Z_3]$ of finite length. This is obvious if one observes that

$$\dim_{Z_2} \operatorname{Ker}(d_s) \equiv (-1)^{s+1} \pmod 3$$

(since $\dim_{Z_2} A = 3$) so that $\operatorname{Ker}(d_s)$ can never be 0.

However, with slight modifications, the algorithm above will produce free resolutions even when A is not connected. The modifications are in the second phase, which should be replaced by

```
For cyc ∈ Oldker begin
    Reduce cyc against Image;
    If cyc ≠ 0 then begin
        add a generator g to Ext^s;
        diff(g) = cyc;
        deg(g) = t;
        insert (g,cyc) into Image;
        For op ∈ Basis of A_0 begin
            x := op * g;
            dx := Act(op,cyc);
            Reduce (x,dx) against Image;
            If dx = 0 then append x to Newker
                      else insert (x,dx) into Image;
        end {op};
    end {if};
end {cyc};
```

This algorithm is correct, in the sense that it terminates. However, the size of the resolution it produces is sensitive to the order in which the basis for Oldker is processed. Any efficient algorithm for producing a minimal (or at least small) generating set over A for an A-submodule of a free A-module, given a k-basis for it, would be very useful here. I would be happy to know one even in the case $A = Z_2[\Sigma_n]$ for n = 5 or 6. One obvious first step in this direction would be to order the k generators of Oldker by the dimension of the A_0-submodule of the quotient Oldker/Image that they generate and hit one of maximal dimension first. This would alter the dimensions for the remaining elements, however, so could be a very slow process.

In the nonconnected case, the free resolution requires further processing in order to produce Ext. Of course, we will actually compute Tor and use duality to compute Ext. To compute Tor, we replace each $\operatorname{diff}(g) = \sum op_i * g_i$ by $\sum \varepsilon(op_i) * g_i$, where $\varepsilon : A \to k$ is the augmentation. We then recompute Image and Newker and the cokernel

$$\text{Image} \to \text{Oldker} \to \text{Tor} \to 0$$

just as before. This second homology computation is very fast compared to the first, since $\dim_{Z_2}(D \otimes_A Z_2) = \dim_A(D)$ is generally much smaller than $\dim_{Z_2}(D)$.

We will not discuss the nonconnected case any further in this chapter.

3. PRODUCTS

There are two sorts of products in Ext which we could use to make H(A) an algebra and H(M) an H(A)-module. One is the external product

$$\operatorname{Ext}_A(M,k) \otimes \operatorname{Ext}_A(k,k) \to \operatorname{Ext}_{A\otimes A}(M,k)$$

induced by the tensor product of modules, and its internalization by pullback along the coproduct $A \to A \otimes A$, when A is a Hopf algebra. However, this is of little help when A is not a Hopf algebra. Also, it presents computational problems because of the size of the tensor product complex: if C and D strain memory capacity separately, then $C \otimes D$ is completely out of reach. The other possibility, Yoneda's composition product

$$\operatorname{Ext}_A(M,k) \otimes \operatorname{Ext}_A(k,k) \to \operatorname{Ext}_A(M,k)$$

suffers neither of these restrictions. In fact, it is vastly more efficient in memory use, since for long stretches of the computation, only a single filtration of each of C and D is needed. Thus, it is efficient to keep in memory only those differentials for the filtrations currently of interest. Another advantage is that we may easily restrict attention to the action on indecomposables using the Yoneda definition, a task which is less naturally accomplished using the tensor product.

In fact, there are two versions of Yoneda's composition product. The one we will _not_ use views Ext as equivalence classes of extensions. The one we want uses the isomorphism between $\operatorname{Ext}^s(M,k)$ and the chain homotopy classes of degree s chain maps from D to C, where D and C are free resolutions of M and k, respectively.

In these terms, the product is just composition of chain maps. In terms of representative cocycles, $[y][x] = [y x_{s+s'}]$, where $[x]$ denotes the cohomology class of the cocycle x.

$$\begin{array}{ccccccccccc} 0 & \leftarrow & M & \leftarrow & D_0 & \leftarrow \cdots \leftarrow & D_s & \leftarrow & \cdots & \leftarrow & D_{s+s'} \\ & & & & & \swarrow^{x} & \downarrow \tilde{x}_s & & & & \downarrow \tilde{x}_{s+s'} \\ & & & & k & \leftarrow & C_0 & \leftarrow & \cdots & \leftarrow & C_{s'} \\ & & & & & & & & & & \downarrow y \\ & & & & & & & & & & k \end{array}$$

In the following lemma we translate this into terms suitable for mechanical calculation. If $g \in D_s$, let $g^* : D_s \to k$ be the cochain dual to g with respect to a fixed k-basis $\{a_i g_j\}$ of D, where $\{g_j\}$ is an A-basis of D and $\{a_i\}$ is a k-basis of A. If $y : D_s \to k$ is a cocycle (e.g., $y = g_i^*$ for some i), then it can be lifted to a chain map $\tilde{y} : D \to C$.

Lemma 3.1

If $\{h_i\}$ and $\{g_i\}$ are A-bases of the minimal free resolutions C and D, then

$$h_i^* g_j^* = \sum_g h_i^*(\widetilde{g_j^*}(g))g^*$$

summing over all g of the correct homological degree. That is, the coefficient of g^* in $h_i^* g_j^*$ is the coefficient of h_i in $\widetilde{g_j^*}(g)$.

Thus, the entire H(A) action on the element g_j^* can be seen by inspecting the chain map $\widetilde{g_j^*}$.

In fact, with a bit of care in setting up $d : C_1 \to C_0$, the action of $H^1(A)$ can already be seen in the differential of D. Let I be the augmentation ideal of A, and choose $\{a_i\} \subset I$ so that $\{a_i + I^2\}$ is a k-basis for I/I^2. Let $\{a_i'\}$ be a k-basis for I^2. Then let us take $\{1\} \cup \{a_i\} \cup \{a_i'\}$ as our k-basis of A. Since C is minimal, $C_0 = A$ generated by 1, and C_1 is free over A on a set $\{h_i\} \cong \{a_i\}$. We may assume that $d(h_i) \equiv a_i \pmod{I^2}$.

Lemma 3.2

$h_i^* x^* = (a_i x)^* d$. That is, the coefficient of g^* in $h_i^* x^*$ is the coefficient of $a_i x$ in $d(g)$.

Proof. The k-linear (but not A-linear) maps a_i^* and $(a_i x)^*$ satisfy $h_i^* = a_i^* d$ and $a_i^* (\widetilde{x^*})_s = (a_i x)^*$. Therefore,

$$
\begin{aligned}
h_i^* x^* &= h_i^* (\widetilde{x^*})_{s+1} \\
&= a_i^* d(\widetilde{x^*})_{s+1} \\
&= a_i^* (\widetilde{x^*})_s d \\
&= (a_i x)^* d
\end{aligned}
$$

We compute the chain map $\widetilde{g}^*$ induced by a cocycle g^* in exactly the same way we would prove such a lift exists. The algorithm follows.

```
(Let g ∈ Ext^{s't'}.)
For g1 ∈ Ext^{s'}
    if g1 = g then g̃*(g1) := 1
              else g̃*(g1) := 0;
For s = 1 to ∞
    For t = conn(D_{s+s'}) to maxt(s) begin
        compute Image(d : C_{st} → C_{s-1,t});
        For g1 ∈ Ext^{s+s',t+t'} begin
            x := 0;
            dx := g̃*(diff(g1));
            Reduce (x,dx) against Image;
            g̃*(g1) := x;
        end {g1};
    end {t};
end {s}.
```

To compute the image, we use the same loop as in the homology program, so we have abbreviated it here. Note that in order to compute $\widetilde{g}^*$ on D_{st}, we must have already computed it on $D_{s-1,t}$ and hence on D_{s-1,t_1} for $t_1 \leq t$. Any ordering of the computation which ensures this will be correct. However, the order we have used has the advantage that for each s, the only differentials needed are those on $D_{s+s'}$

and C_s, and the only values of $\tilde{g}^*$ needed are those on D_{s-1}. This cuts memory requirements at a very minor I/O cost: we read in the differentials we need at the beginning of each s loop and purge the values no longer needed at the end. In the homology program, the fact that we also have to compute the kernel of the differential forces us to vary s faster than t, making this kind of saving impossible.

4. TODA BRACKETS

If we compute products as Yoneda composites of chain maps, it is natural to replace Massey products by Toda brackets. To get the signs right with a minimum of clutter, we find the following conjugation operation useful.

Definition 4.1

If $c : W \to X$ is a degree s homomorphism of graded A-modules ($c_i : W_i \to X_{i-s}$), let $\bar{c} : W \to X$ be the homomorphism with ith component $(\bar{c})_i = (-1)^i c_i$.

Clearly, conjugation is linear in c and is its own inverse. Also, one can quickly verify that

$$\overline{ab} = a\bar{b} = (-1)^{\deg(b)}\bar{a}b$$

Thus, a is a chain map iff its conjugate is an "anti-chain map":

$$da = ad \Longleftrightarrow d\bar{a} = -\bar{a}d$$

Similarly, for a chain null homotopy $x : a \simeq 0$,

$$dx + xd = a \Longleftrightarrow d\bar{x} - \bar{x}d = \bar{a}$$

It follows that two null homotopies of the same map differ by the conjugate of a chain map:

$$dx + xd = dy + yd \Longleftrightarrow d\overline{(x - y)} = \overline{(x - y)}d$$

Definition 4.2

The suspension ΣW of a chain complex W has $(\Sigma W)_i = W_{i-1}$ and the same differential as W. The mapping cylinder C(c) of a chain map $c : W \to X$

of degree s is the complex with ith component $C(c)_i = X_i \oplus W_{i-s-1}$ and differential

$$\begin{pmatrix} d & \bar{c} \\ 0 & d \end{pmatrix}$$

The natural inclusion $i : X \to C(c)$ and projection $\pi : C(c) \to \Sigma W$ are chain maps, and the following is a cofibration sequence.

$$W \xrightarrow{c} X \xrightarrow{i} C(c) \xrightarrow{\pi} \Sigma W$$

Definition 4.3

If $W \xrightarrow{c} X \xrightarrow{b} Y \xrightarrow{a} Z$ are chain maps, the Toda bracket $\langle a,b,c\rangle$ is the set of all chain maps T such that the following diagram homotopy commutes for some chain map H.

$$\begin{array}{ccccccc} W & \xrightarrow{c} & X & \xrightarrow{b} & Y & \xrightarrow{a} & Z \\ & & & \searrow^{i} & \uparrow H & & \uparrow T \\ & & & & C(c) & \xrightarrow{\pi} & \Sigma W \end{array}$$

Proposition 4.4

1. $\langle a,b,c\rangle = \{T \mid T \simeq a\bar{y} - x\bar{c}$ where $b' \simeq b$, $x : ab' \simeq 0$, and $y : b'c \simeq 0\}$.
2. $\langle a,b,c\rangle = \{T \mid T \simeq a\bar{y} - x\bar{c}$ where $x : ab \simeq 0$ and $y : bc \simeq 0\}$.
3. $\langle a,b,c\rangle$ depends only on the homotopy classes of a,b, and c.
4. Up to chain homotopy, the indeterminacy $\{f - g \mid f,g \in \langle a,b,c\rangle\}$ is $a[W,Y] + [X,Z]c$, where $[-,-]$ denotes the set of chain homomorphisms.

<u>Proof</u>.

1. First, suppose given chain maps H and T, and homotopies $\lambda : Hi \simeq b$ and $\mu : T\pi \simeq aH$. Write $H = (b',\bar{y})$ and $\mu = (-x,\phi)$. Then we find that $\lambda : b' \simeq b$, that

   $$H \text{ is a chain map} \iff b' \text{ is a chain map and } y : b'c \simeq 0$$

   and that

   $$\mu : T\pi \simeq aH \iff x : ab' \simeq 0 \text{ and } \phi : T \simeq a\bar{y} - x\bar{c}$$

Conversely, if $\lambda : b' \simeq b$ and $y : b'c \simeq 0$ then $H = (b', \bar{y})$ is a chain map and $\lambda : Hi \simeq b$. If, in addition, $x : ab' \simeq 0$ then $a\bar{y} - x\bar{c}$ is a chain map. Finally, if $\phi : T \simeq a\bar{y} - x\bar{c}$, then $(-x, \phi) : T\pi \simeq aH$.

2. Suppose given λ, x, and y as in 1. Then $y - \lambda c : bc \simeq 0$ and $x - a\lambda : ab \simeq 0$. Thus, $a(\overline{y - \lambda c}) - (x - a\lambda)\bar{c}$ is in the right-hand side of 2. But $a\bar{y} - x\bar{c} = a(\overline{y - \lambda c}) - (x - a\lambda)\bar{c}$, so 1 and 2 agree.
3. For b, this follows from 1. For a, suppose $\lambda : a' \simeq a$, $x : ab \simeq 0$, and $y : bc \simeq 0$. Then $x + \lambda b : a'b \simeq 0$, so

$$a'\bar{y} - (x + \lambda b)\bar{c} \in \langle a', b, c\rangle \qquad \text{and} \qquad a\bar{y} - x\bar{c} \in \langle a, b, c\rangle$$

and the map $\lambda\bar{y}$ is a chain homotopy between these. Similarly, if $\lambda : c' \simeq c$, then $x\bar{\lambda}$ is the chain homotopy we need.
4. By 2 we may assume $f \simeq a\bar{y} - x\bar{c}$ and $g \simeq a\bar{y_1} - x_1\bar{c}$, where x and x_1 are null homotopies of ab, and y and y_1 are null homotopies of bc. Then

$$\begin{aligned} f - g &\simeq a(\bar{y} - \bar{y_1}) - (x - x_1)\bar{c} \\ &= a(\overline{y - y_1}) + (-1)^{\deg(c)}(\overline{x - x_1})c \end{aligned}$$

which, by the observation preceding Definition 4.2, has the form claimed.

This description is rather extravagant computationally. We do not need the entire chain map defining the Toda bracket, only the cocycle it lifts. This is expressed by the following corollary.

Corollary 4.5

Let a be a cocycle and $\tilde{a} : Y \to Z$ a chain map covering a. Let $W \xrightarrow{c} X \xrightarrow{b} Y$ be chain maps. If $y : bc \simeq 0$ then the map $a\bar{y}$ is a cocycle representing an element of $\langle \tilde{a}, b, c\rangle$. Thus, the coefficient of g^* in $\langle \tilde{a}, b, c\rangle$ is the coefficient of a in $\bar{y}(g)$.

Proof. The component of the null homotopy $x : \tilde{a}b \simeq 0$ which maps into Z_0 is 0. Therefore, the cocycle corresponding to the chain map $\tilde{a}\bar{y} - x\bar{c}$ is $a\bar{y}$.

Thus, computing a null homotopy of bc gives us all Toda brackets of the form $\langle a,b,c\rangle$. This is particularly useful in computing periodicity operators in the cohomology of the Steenrod algebra and its subalgebras since they can be expressed as

$$P^n(x) = \langle x, h_0^{2n}, h_n\rangle$$

At first glance, this corollary appears to say that the null homotopy of $\tilde{a}b$ is irrelevant and only that of bc matters. However, the asymmetry is more apparent than real. The homomorphism $\tilde{a}\bar{y}$ is not a chain map, and in order to lift the cocycle $a\bar{y}$ to a chain map we have to subtract $x\bar{c}$ from $\tilde{a}\bar{y}$. Thus, the complementary term $x\bar{c}$ is implicit in the cocycle which we have shown represents the Toda bracket.

As in the case of products, elements of $H^1(A)$ have special behavior which reduces the amount of computation needed to find Toda brackets involving them. Retain the assumptions made preceding Lemma 3.2 about $d: C_1 \to C_0$ and recall that $d(h_i) \equiv a_i \pmod{I^2}$.

Corollary 4.6

$\langle \tilde{h}_i^*, \tilde{b}^*, c\rangle = (a_i b)^* \bar{c}$. That is, the coefficient of g^* in $\langle \tilde{h}_i^*, \tilde{b}^*, c\rangle$ is the coefficient of $a_i b$ in $\bar{c}(g)$.

Proof. We use the preceding corollary and the same k-linear maps as in the corresponding result for products:

$$h_i^* \bar{y} = a_i^* d\bar{y} = a_i^* b^* \bar{c} = (a_i b)^* \bar{c}$$

where the middle equality follows from the general formula $d\bar{y} = \bar{y}d + b\bar{c}$ and the fact that the component of $\bar{y}$ mapping into C_0 is 0.

The algorithm for calculating a null homotopy y of $b^* c^*$ is not significantly different from the algorithm for computing an induced chain map. We initialize the calculation by setting $y = 0$ on $D_{s'+s''-1}$, where $s' = \deg(b)$ and $s'' = \deg(c)$, and then proceed to lift $b^* c^* - yd$ over the differentials $d: C_{st} \to C_{s-1,t}$.

5. REPRESENTATION OF THE ALGEBRA

We assume we are given an ordered k basis $\{a_i\}$ for A such that each degree is linearly ordered (and hence well ordered, since A has finite type). In terms of this basis, there are two natural representations for the elements of A:

Sparse case: an ordered list $((k_{i_1}\ a_{i_1})(k_{i_2}\ a_{i_2})\ \cdots)$ of the nonzero terms

Dense case: a vector $(k_1\ k_2\ \cdots)$ of the coefficients for all the basis elements in the degree under consideration

Whichever of these representations we use, the operations that we need to perform are:

1. Produce the basis for a given degree.
2. Compare elements to determine which has the smaller leading term.
3. Add and subtract elements.
4. Multiply elements.

We find it convenient to deal with the two cases (sparse and dense) separately. We have used both in our calculations involving the mod 2 Steenrod algebra and have found the density of nonzero coefficients in the minimal resolutions we have constructed such that the dense representation uses roughly one-fourth the storage of the sparse representation. In general, the dense case representation is more compact iff $jd < CN$, where

j is the number of bits per coefficient in the dense representation.

d is $\dim_k A_n$, where n is the degree in question.

C is the average number of nonzero coefficients in the algebra elements under consideration.

N is the number of bits required to hold a coefficient and an identifier for a basis element of A_n in the sparse case.

Clearly, C will be difficult to compute exactly and will have to be estimated. Also, the sparse representation can be made more efficient in its use of space by reducing N, but this may make everything else more difficult.

5.1 The Dense Case

It is trivial to produce the vector representations of the basis elements if we know the dimension of the algebra in the degree of interest. They are (1,0,0,...), (0,1,0,...), etc.

Comparison for order is similarly easy, since the leading term is the first one with a nonzero coefficient. In our current implementation, we have an order-preserving function which assigns to an element an integer identifying the first nonzero coefficient. With each entry (a, da) in Image we also store the integer identifying the leading coefficient of da. Then Reduce has only to search dx to determine whether to replace (x,dx) by (x − a,dx − da), to look at the next entry in Image, or to terminate the search. Note that after replacing (x,dx) by (x − a,dx − da), the leading operation increases, so the search for it can start at the old leading operation rather than at the beginning.

Addition (and subtraction) are somewhat different for $k = Z_2$ and for $k = Z_p$, $p > 2$. When $p = 2$, the vector of coefficients is a sequence of bits, and addition is exclusive-or of these bitstrings. This can generally be accomplished in comparatively few instructions even for long bitstrings. When $p > 2$, if we allow each coefficient one bit more than it must have, i.e., n bits, where $2^{n-2} < p < 2^{n-1}$, then we may pack these into words (16, 32, or 64 bits typically) and add without danger of "interference" between adjacent coefficients. Adding $(2^{n-1} - p, 2^{n-1} - p, \ldots)$ then causes the leftmost bit to reflect the need (or lack of it) for reduction mod p in that coefficient, which may then be carried out a word at a time by a short sequence of logical and arithmetic operations. (I owe this observation to David Anick.)

Multiplication is the most difficult operation in this representation. Note that, aside from needing to know $\dim_k A_n$ for each n, the first three operations are perfectly generic. The multiplication routine is the opposite: it is completely specific to the algebra. In order to implement it, we find that we alter the way in which the first three operations are carried out somewhat. By linearity, it is sufficient to be able to multiply basis elements. Typically, to multiply a_i and a_j, given only i and j, we must first produce some more intrinsic representation of a_i and a_j than their sequence numbers i and j. We then apply the multiplication routine specific to the algebra to those intrinsic representations. We generally receive the answer in intrinsic terms, which must then be converted back into sequence numbers. In principle, then, we must be able to compute Opno(i), the intrinsic representation of a_i, and Seqno(op), the sequence number of the intrinsic representation op. In practice, a_i and a_j are not randomly distributed, so we take a slightly different approach. In the cohomology program, a_i is being stepped through the entire basis for its degree, one element at a time, while a_j runs through the nonzero terms of some element. In the products and Toda brackets programs, both a_i and a_j run through the nonzero terms of an element. Thus the functions we use are

Firstop(n), which produces the intrinsic representation of the first operation of degree n

Nextop(op), which produces the intrinsic form of the successor to op

Advance(op,k), which produces the intrinsic form of the operation k steps beyond op

We use Firstop and Nextop to produce each basis element in a given degree. We use Advance to skip through an element's nonzero terms, since it is much faster to produce the intrinsic form of a term from that of the preceding nonzero term than to produce it from the sequence number. Depending on the algebra, we may define one of Nextop and Advance in terms of the other.

For the transformation from intrinsic form to sequence number, we use

Seqno(op) = sequence number of the basis element with intrinsic representation op

$\text{Rseqno}(op_1, op_2) = \text{Seqno}(op_1) - \text{Seqno}(op_2)$

(Of course, for efficiency, we probably will not compute Rseqno by this formula.) The point of Rseqno is that, for the Steenrod algebra at least and certainly in many other cases, it is possible to devise the multiplication routine so that it produces the terms in its answer in roughly ascending order. If we already know the sequence number of the preceding operation, it may be quite a bit faster to compute only the relative sequence number of the next term and add, rather than compute its sequence number from scratch.

5.2 The Sparse Case

Here we must assign an identifier, which we take to be an integer, to each a_i. It is convenient to use a "packed" version of the intrinsic form for speed of conversion in the multiplication routine.

We order $\{a_i\}$ by numerical order of their identifiers. Of course, we will try to arrange the packed version so that this coincides with an ordering that makes sense for the algebra. For example, with the mod 2 Steenrod algebra's monomial basis $\{\xi_1^{r_1}\xi_2^{r_2}\cdots\xi_n^{r_n}\}$, we use the identifier

$$\mathrm{Pack}(r_1,r_2,\ldots,r_n) = r_1 + 2^8(r_2 + 2^7(r_3 + 2^6(r_4 + 2^4(r_5 + 2^3(r_6 + 2^2r_7)))))$$

which is easily computed by a sequence of shift operations. Numerical order of the packed forms then coincides with reverse lexicographic ordering of the monomials based on the ordering $\xi_1 < \xi_2 < \cdots$ of generators. Note that this function is only one-to-one through dimension 240, so will have to be replaced if we ever exceed that dimension.

Of course, we also need the inverse function Unpack, which takes an integer identifier and produces the corresponding intrinsic form of the basis element.

To produce the basis for a given degree, the operations Firstop(n) and Nextop(op) composed with Pack will work fine. Note that Nextop must return some indication when its argument is the last basis element in its degree.

Comparison of elements is just lexicographic order of lists based on the standard order for the integers:

$$(i_1 i_2 \cdots) < (j_1 j_2 \cdots) \Longleftrightarrow i_1 < j_1 \text{ or } (i_1 = j_1 \text{ and } (i_2 \cdots) < (j_2 \cdots))$$

Addition (or subtraction) is a kind of merge with cancellations, since the lists are presumed ordered: we perform a merge of the identifiers i_j and i'_j of the lists

$$((k_1\, i_1)(k_2\, i_2)\cdots) \qquad \text{and} \qquad ((k'_1\, i'_1)(k'_2\, i'_2)\cdots)$$

where k_i, $k'_i \in Z_p$ and $i_1 < i_2 < \cdots$ and $i'_1 < i'_2 < \cdots$, adding coefficients when the same identifier occurs in both and eliminating the term if the sum of the coefficients is 0. When $k = Z_2$ we omit the coefficients k_i, naturally, and eliminate terms which occur in both lists (we're computing the symmetric difference of sets in this case).

Multiplication is slightly simpler than in the dense case. We simply pass through the list, unpacking each item in order to multiply. The basis elements making up the product are packed and inserted in order into the list making up the sum so far. Note that the efficiency of this last insertion is enhanced if we can arrange the multiplication routine to produce its results in order, or in reverse order, depending on the precise method we use to store lists. If we use a LISP-style arrangement

then reverse order is best, since each insertion requires only one comparison to verify that it goes at the front. If we use contiguous allocation

i_1
i_2
$\vdots$
i_n
$\vdots$

increasing order is probably best, because insertion at the end is most efficient. (Insertion anywhere else requires copying whatever follows.)

5.3 Polynomial and Copolynomial Algebras

Note that if A or the dual of A is a polynomial algebra, then the manipulation of the basis elements consists of standard operations on weighted partitions. In particular, this is the case for the Steenrod algebra and its subalgebras.

6. REPRESENTATION OF MODULES

6.1 Free Modules

The modules which make up the resolution are free, so are especially easy to represent. If $\{h_i\}$ is a well-ordered A-basis for the free A-module C, then we represent elements of C by ordered lists $((op_1\ h_{i_1})$ $(op_2\ h_{i_2})\ \cdots)$, where $h_{i_1} < h_{i_2} < \cdots$ and the op_j are algebra elements (not just basis elements) represented as in the previous section. Such a list represents the obvious sum $\sum op_j h_{i_j}$.

Addition of such elements is essentially the same as addition of algebra elements in the sparse representation. To act on such an element by an element of the algebra is just the obvious bilinear extension of the multiplication of algebra elements.

6.2 General Modules

The module being resolved is defined to the programs by two routines: one writes a k-basis for a specified degree of the module on the file which holds Oldker, and the other returns the result of letting a basis element of the algebra act on an element of the module.

We represent <u>all</u> module elements in the same format as free module elements since this means that the same addition routines and I/O routines can be used. Conceptually, this amounts to representing a module as a quotient of a free module. The routine which gives a k-basis for the module is equivalent to giving a k splitting of the quotient homomorphism.

A general technique is to represent the module M as the quotient of the free A-module on a k basis $\{m_i\}$ of M. Each module element then has the form $((1\ m_{i_1})\ (1\ m_{i_2})\ \cdots)$, where 1 is the identity element of A. To define the action of A on M then requires that we specify the elements $a_i m_j$ in this form, for each a_i in the k basis of A.

6.3 Special Cyclic Modules

Cyclic modules which are at most one-dimensional (over k) in each degree can be represented in a much simpler fashion, as quotients of A. If m is the generator of the module, then to define the k basis for degree t + deg(m) of the module, we need only specify a basis element a_{i_t} of A such that $a_{i_t} m \neq 0$. To define the action of A on M we give, for each degree t, the projection $A_t \to k$ defined by letting A_t act on m and recording the coefficient of the result. When $k = Z_2$ this is a bitstring of length $\dim_k A_t$. Then, to compute a_i acting on ((op m)) we compute the product $a_i * op$ and apply the projection. When $k = Z_2$, this consists of a logical-and of the bitstring representing the product and the bitstring representing the projection, followed by counting (mod 2) the number of nonzero bits. The result is then the result of the projection times $((a_{i_{t'}}\ m))$, where $t' = \deg(op) + \deg(a_i)$.

7. DATA STRUCTURES

We have found that a single datatype, which we call a block, suffices for all our dynamically allocated memory. It consists of an even number, say 2n, of contiguous 4-byte words, which we think of as split into two half-words (2 bytes each), followed by 2n − 1 full words. The first two bytes contain either 2n − 1 or 2n − 2, reflecting the number of words of data following the first word of the block. The second two bytes of the block (i.e., the last half of the first word) also hold data, generally of a different nature than the data held in the full words. We call these two bytes the <u>key</u>, and the first two bytes the <u>length</u>.

$L = 2n-2$

	block	
	L | *key*	1
1		2
⋮	⋮	⋮
L		
	unused	2n

or

$L = 2n-1$

	block	
	L | *key*	1
1		2
⋮	⋮	⋮
L		2n

The use of these blocks to represent lists has the advantage of keeping elements of a list contiguous, speeding access (via binary search or hashing), and allowing the use of machine operations which act on large blocks of memory. Of course, it also introduces the need for memory management, which we discuss below, and means that inserting elements into the middle of lists requires some copying.

We use a single block to represent an element of the algebra, in either the sparse form or the dense form. In the sparse form, each word of the block holds an identifier for a basis element. In the dense form, the words in the block are thought of as one long string of bits. In either case the key field is not used by the algebra element. Instead, it is used to hold an identifier for a module generator, so that a block can represent a term of the form op * gen.

To represent an element of a module, we use a block whose entries are themselves (pointers to) blocks representing terms op * gen as just described. The key of the main block is set equal to the internal degree of the module element. Thus, the structure

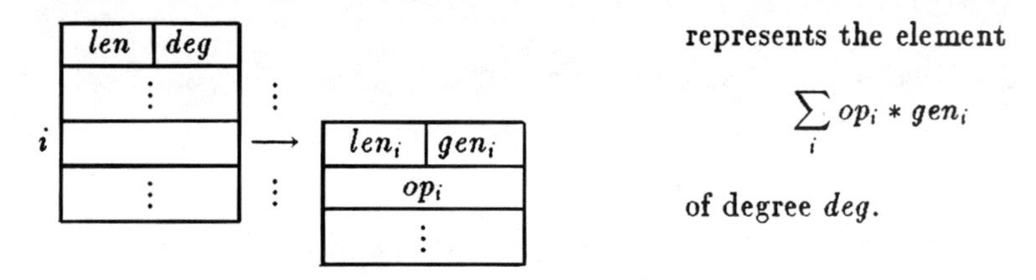

The lists Ext^s of generators for the resolution in cohomological degree s are examples of lists which need to be able to grow. To keep such lists in a single block without having to recopy them each time an element is added, we use a block whose length field is the full length $2n - 1$ of the block, while the key field is set to the actual data length of the block. When the block is full, we copy it into a new block with some room to spare for expansion.

We store Image as a block whose ith entry is itself a block holding the list of (a,da) entries with leading generator of da equal to the ith generator in Ext^{s-1}. Thus, Image is a block whose length is the same as the data length of Ext^{s-1}. When we are reducing (x,dx), we find the leading generator gen of dx in Ext^{s-1} by binary search and look at the block located at the corresponding position in Image. This block is an expandable block, each entry of which corresponds to an (a,da) pair, with the entries arranged in increasing order of the leading operation lop on gen. Thus, another binary search determines whether or not Image contains an (a,da) pair whose leading term lop * gen agrees with the leading term of dx. Each (a,da) pair is stored in a block of length 3, along with the disk address addr at which a and da were written and an identifier lop for the leading operation in da, so that we may compare it to the leading operation of dx without having to look at da. This is important primarily if the pair (a,da) has been paged out, since it means we may determine whether or not we need (a,da) to reduce (x,dx) without having to read (a,da) back in. Thus, we will not read it back in unless we actually need it.

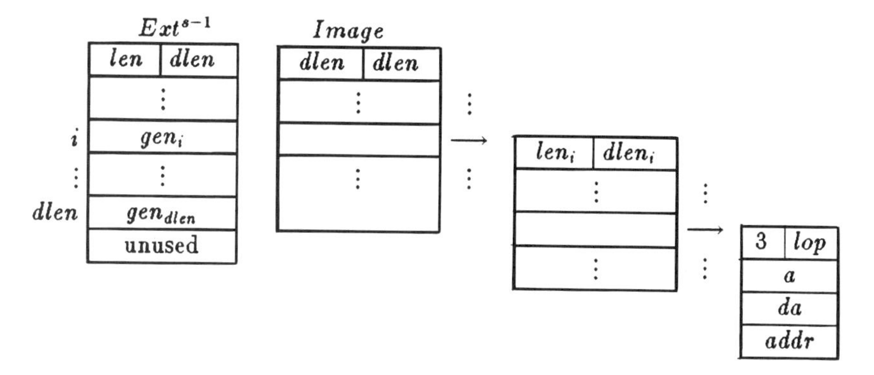

Organizing Image in this way has two significant beneficial effects. First, searching Image for an (a,da) pair with leading term the same as the leading term of (x,dx) is faster, because when one is found, and (x,dx) is replaced by (x − a,dx − da), we can continue our search inside the subblock if the leading generator is not changed, and we can immediately skip the rest of the subblock if it is. Second, when a new (a,da) entry is to be inserted, only the part of the subblock following it has to be moved down. This can significantly reduce the amount of copying necessary.

The only other dynamically allocated data structures needed are Newker and Oldker. Since these are generated and used in one sequential pass, they are simply written on and read from sequential files.

8. MEMORY MANAGEMENT

We maintain a free list of unused blocks in order of address. This permits amalgamation of adjacent free blocks. We allocate space from this list by first fit; that is, the block requested is carved from the first block large enough to hold it and the remainder (if any) is left in place. An exact length or a range of acceptable lengths can be requested. The latter form is used for lists that will expand (such as

Ext^S) and for blocks that are likely to soon have a piece removed from the end. For example, in computing the sum of elements of lengths i and j, we request a block of length i + j to hold the sum. However, cancellations may mean that the sum will have fewer than i + j entries and the remainder of the block will have to be freed. Thus, if a block of length i + j + 2 is available, it makes more sense to take all of it than to chop off two words now, then perhaps chop off a few more momentarily, especially since they will likely have to be reattached to the two originally removed.

A block in the free list must contain at least two words in order to hold its length information and the address of its successor. Blocks are allocated in multiples of 2 words to ensure this.

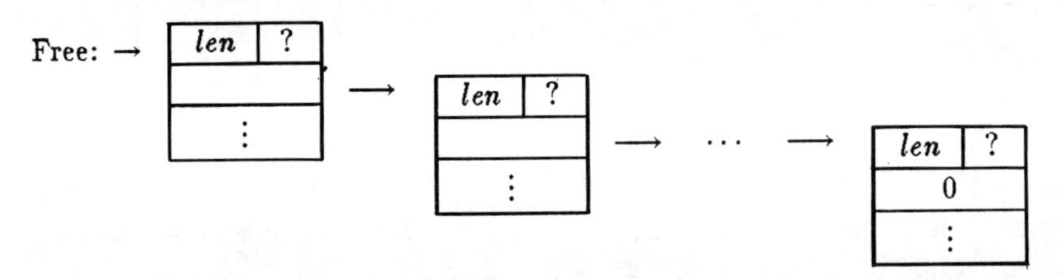

If less than an entire block is being returned, it is taken from the end of the first large enough block. In this way, no pointers in the free list have to be changed.

Our original memory allocation system also maintained separate lists of common size blocks. This eliminated the need to search the free list when a block of such a size was requested and eliminated the need to chop larger blocks into pieces. However, it increased fragmentation to such an extent that it was counterproductive. (Segregating the free blocks into separate lists of different sizes eliminated much of the amalgamation that takes place when all the blocks are in the same free list.)

To improve locality, a region called the <u>clear</u> area is maintained. Initially, the free list is empty and the clear area consists of all space allocated for blocks. If a request for a new block cannot be granted from the free list, the allocation routine checks to see if it can be granted from the clear area. If so, the requested amount of space is taken from there. Similarly, if a block located on the edge of the clear area is freed, the clear area is expanded to include it, instead of adding it to the free list. A similar effect could be achieved by taking small blocks from the beginning rather than the end of oversize blocks, or by ordering the free list in reverse order. However, with only 16 bits for the length field of the block, we are limited to blocks whose length is 64K words, quite a bit less than the initial size of memory requested for some computations, which would force us to begin by carving it into pieces. If memory were requested from the operating system in 64K word (or smaller) segments, this latter method of encouraging locality would be the method of choice.

When space for a new block is requested and no space is available, we begin paging out parts of Image, since this is what uses most of memory. The scheme we have adopted is rather crude but effective. We remove from memory the last one-fourth of the entries in each of Image's subblocks. More precisely, we remove a and da but keep the block containing lop, addr, and space to put pointers to a and da if they are returned to memory. If that is sufficient, we stop there. If not, we remove the one-fourth before that, etc. An entry is returned to memory only when it is again needed. This works as well as it does largely because we have also arranged to produce elements in reverse order, so that their position in Image can be rapidly established. We should also point out that when we remove an entry from memory, no I/O is required, because every entry is written out as it is generated and will never change. Thus, only half of the I/O cost commonly associated with paging is incurred.

Finally, if the algebra is actually finite-dimensional, not just of finite type, we remove differentials that can no longer contribute to the calculation. (If we have reached internal degree t and the algebra is 0 above degree t', then differentials in degrees less than $t - t'$ can have no further effect.)

9. I/O

Module elements have a linear form which consists of the first word of their main block, containing their length and degree, followed by the contents of their subblocks, in order. When writing a block of odd total length, i.e., even data length, the unused last word is omitted. Writing this to a file is elementary. Similarly, reading it is easy. The first word tells us the size block to allocate and the number of further blocks to read. For each of those, the first word we read tells us how many more words we need to read.

The files holding Oldker and Newker consist of a header identifying the bidegree for which it is the kernel, following by a stream of elements spanning the kernels linearized in the above fashion. The end is indicated by an element of length 0. At the end of the computation for each bidegree the files holding Oldker and Newker swap roles.

The other files we use are all random access files. These are

Diff, which holds the differentials defining a resolution

Map, which holds the chain maps lifting cocycles corresponding to a set of generators of a particular homological degree

Image, which holds the image of the differential $d : C_{st} \to C_{s-1,t}$ while that image is in use

Each of these contains a header that tells

The number of words in the file

The number of words in the header (allowing for a variable-length description of the file located between this header and the main body)

The last bidegree completed

The module(s) involved

(Optional) text describing the contents of the file

In addition, the header for a Map file gives the number n, homological degree s, and identifiers $g_1, \ldots, g_n$ of the generators of D whose induced chain maps $\tilde{g_i}^* : D_{s+s'} \to C_{s'}$ are contained in the file.

Following the header, the file is simply a sequence of individual entries. In Diff the entries have the form (g,s,t,diff), where g is the identifier for a generator of the resolution, s and t are its homological and internal degrees, and diff is the linearization of the differential on g. In Map the entries have the form (i,g,elt), where elt is the linearization of $\tilde{g_i}^*(g)$. In Image the entries have the form (dx,x), where dx and x are the linearizations of the two elements. We put dx first because in the second phase of the homology calculation the x entries are no longer needed. Thus, if we need to page dx back in, its address is the same as the address of the pair.

The functions needed to carry out the I/O are

Read or Write the linearization of an element at the current position in a sequential (kernel) file.

Rewind a sequential file.

Read or Write the linearization of an element at a specified address in a random-access file, and report the address of the next word following it.

Read or Write a vector of specified length at a specified address in a random access file.

10. RESULTS

The previous version (using the sparse representation of algebra elements) was used to calculate a minimal resolution of the Steenrod algebra through internal degree 69, together with the chain maps induced by indecomposables in homological degrees 3 to 6 through this range.

With the current version of the program (using the dense representation of algebra elements), we have computed

A minimal resolution of the Steenrod algebra through internal degree 55

A minimal resolution of A(2), the subalgebra of the Steenrod algebra generated by Sq^1, Sq^2, and Sq^4, through internal degree 90

Minimal resolutions for $H^*(RP^\infty/RP^{n-1})$ over A(2), for n = 1,3,5,7, through internal degree 45

A minimal resolution for a cyclic A(2) submodule of $H^*(MO\langle 8\rangle)$, through internal degree 80

Chain maps for indecomposables in low homological degrees for the resolutions of A(2) and the submodule of $H^*(MO\langle 8\rangle)$

Our next projects are to extend the calculation of the minimal resolution of the Steenrod algebra and compute all products and many Toda brackets in it, to calculate a minimal resolution of A(3), and to calculate minimal resolutions for some modules over the Steenrod algebra of interest in our other work.

ACKNOWLEDGMENTS

Professor Neil Rickert was helpful in innumerable ways, both in dealing with the MVS and CMS systems and in suggesting and discussing ways to make the algorithms and data organization more efficient. This work was also aided by the excellent computer facilities and staff at the University of Illinois at Chicago. In particular, Dr. Nora Sabelli was always quick to solve problems that arose. I am also grateful to the Research Office and the Computer Center at Wayne State University for funding.

5

EHP Computations of $E_2(S^n)$

EDWARD CURTIS

University of Washington
Seattle, Washington

MARK MAHOWALD

Northwestern University
Evanston, Illinois

1. INTRODUCTION

The E_2 terms of the unstable Adams spectral sequences for spheres may be calculated inductively by an EHP process, as was done (for a range of stem dimensions) in [8,14,19]. In this approach, the main difficulty is computing the homomorphism P. In [8], P was computed by a program called difftag. The purpose of this chapter is to show how this program works and to describe some improvements that have been made. This is to be considered as a sequel to [8]; the notation and conventions are all taken from that paper. All the spaces and groups are localized at the prime 2.

The programs described in [8] were used to compute the groups $E_2^{*,*}(S^n)$ through stem 51. Using the modifications described below, we have continued these calculations through stem 58. It would take up too much space to print the results of these calculations for all spheres; instead, we take this opportunity to give the table for S^3 through stem 58. In a sequel, these results will be used to compute (a part of) the E_∞ term for S^3.

2. EHP SEQUENCES OF UNSTABLE ADAMS E_2 TERMS

For each n, there is a long exact sequence

$$\cdots \to E_2^{*,*}(S^{2n+1}) \overset{E}{\to} E_2^{*,*}(S^n) \overset{H}{\to} E_2^{*,*}(S^{n+1}) \overset{P}{\to} E_2^{*,*}(S^{2n+1}) \to \cdots \quad [2.1]$$

These LESs come about as follows. In [4] it was shown that $E_2^{s,t}(S^n)$ is isomorphic to the homology of a differential module $\Lambda(n)$, obtained as a submodule of the lambda algebra Λ. Λ is (defined as) the algebra (over Z/2) with a generator λ_i for each integer $i \geq 0$, and relations: whenever $2i < j$,

$$\lambda_i\lambda_j = \sum_{k \geq 0} C(j - 2i - 2 - k,k)\lambda_{j-i-k-1}\lambda_{2i+k+1} \qquad [2.2]$$

Then Λ becomes a differential algebra, with differential

$$d(\lambda_i) = \sum_{k \geq 1} C(i - k,k)\lambda_{i-k}\lambda_{k-1} \qquad [2.3]$$

Here $C(n,q)$ stands for the binomial coefficient reduced mod 2.

For each sequence $I = (i_1,i_2,\ldots,i_s)$ of nonnegative integers, I denotes the product $\lambda_{i_1}\lambda_{i_2}\cdots\lambda_{i_s}$. A sequence I is called admissible if for each j, $2i_j \geq i_{j+1}$. It follows immediately from the relations that Λ has for basis (over Z/2) the set of all monomials I, where I is admissible. Λ is bigraded by filtration (length) and dimension, where

$$\text{filtration}(I) = s$$
$$\text{dimension}(I) = i_1 + 1_2 + \cdots + i_s$$

For each positive integer n, $\Lambda(n)$ is defined to be the submodule of Λ spanned by those I which are admissible and for which $i_1 < n$. One of the main results of [4] is that

$$E_2^{*,*}(S^n) = H_*(\Lambda(n))$$

For each n, there is a short exact sequence

$$0 \to \Lambda(n) \overset{i}{\to} \Lambda(n + 1) \overset{h}{\to} \Lambda(2n + 1) \to 0 \qquad [2.4]$$

where i is the inclusion and h is defined (on the basis of admissible monomials) by

$$h(i_1,i_2,i_3,\ldots,i_s) = \begin{cases} (i_2,i_3,\ldots,i_s) & \text{if } i_1 = n \\ 0 & \text{if } i_1 < n \end{cases}$$

The EHP sequence (2.1) for the unstable E_2 terms is the LES in homology of the short exact sequence (2.4). Thus the homomorphism P arises from the differential d in Λ.

3. NOTATION AND CONVENTIONS

Before presenting the algorithm for calculating $E_2^{*,*}(S^n)$, we will briefly describe the relevant techniques from [5,8,14].

3.1 Ordering

As usual, $\Lambda^{s,t}(n)$ denotes the submodule of $\Lambda(n)$ spanned by admissible I of filtration s and dimension t − s. The monomials I of each fixed bidegree (s,t) are ordered lexicographically from the left. This induces a total order on each of the vector spaces $\Lambda^{s,t}$, by first expressing each polynomial as a sum of admissible monomials in decreasing order and then comparing two such polynomials lexicographically. For a sum of admissible monomials, the term which is largest in the lexicographic order is called the leading term. In a given homology class, the polynomial in the class which is least in the total order is called the minimal representative.

If I is the leading term of a minimal representative of some (nonzero) homology class, let c(I) stand for the minimal polynomial which is a cycle and which has I for leading term. We seek a basis of each $E_2^{s,t}(S^n)$ consisting of such basis elements c(I), represented by their leading terms.

As in [8,14], we consider not Λ itself, but the submodule spanned by all λ_I which end with an odd index. The effect of this is that the towers are omitted.

4. THE EHP PROCESS

The EHP process manufactures a file called WFILE as follows. This file consists of a list of rows, each of which is either a sequence of integers (called a term):

$$I = (i_1, i_2, \ldots, i_s)$$

or a pair of such sequences (a term with a tag):

$$(i_1, i_2, \ldots, i_s) \leftarrow (j_1, j_2, \ldots, j_{s-1})$$

A maximum stem dimension N is chosen, and all terms are to have dimension $\leq N$. Initially, WFILE contains all odd positive integers $i \leq N$; each is a term (consisting of a single integer) on a separate row. Then having completed $q - 1$ stages, the qth stage takes three steps, as follows.

1. A program difftag calculates for each J the leading term I of d(J). For each term I of dimension q that is to be tagged by J of dimension $q + 1$, $I \leftarrow J$ is placed in a file called G(q).
2. A program kill reads WFILE and G(q). For each $I \leftarrow J$ in G(q), kill looks for I and J in WFILE; if it finds them both untagged, I is replaced by $I \leftarrow J$ and J is deleted from WFILE.
3. A program loadstem reads WFILE. For each term $I = (i_1, i_2, \ldots, i_s)$ of dimension q, loadstem makes new rows $I' = (i_0, i_1, i_2, \ldots, i_s)$ subject to the conditions that $2i_0 \geq i_1$, and if I is tagged by J, then also $2i_0 < j_1$; these rows are appended to WFILE.

Once the files G(q) are known through some dimension N, the file WFILE, which will contain the computed E_2 terms to stem N, is made by a shell script. This is a sort of master program which calls on the executable programs kill and loadstem to perform steps 2 and 3 for each integer q from 1 to N.

5. THE PROGRAM difftag

The program difftag is more complicated, so we sketch it in a sort of pseudocode.

1. Compute d(J) by formula (2.3) as a linear combination of I's (possibly inadmissible), and place them in a list of rows called LIST; each row is assigned coefficient 1.
2. This LIST is traversed (once) sequentially from the beginning, and each I is tested for inadmissibility from the left. If I is inadmissible at position p, then relations (2.2) are used to express I as a linear combination of terms which are admissible at

position p, and these terms are appended to LIST, each with coefficient 1; the coefficient of I is set to 0.

3. Continue through the list until all terms are admissible (which must occur after a finite number of steps).
4. LIST is searched sequentially, keeping track (by a pointer) of the largest row I with nonzero coefficient. Initially, I is taken as the first term; if a larger term is encountered, the pointer is changed to point to this one. Each later occurrence of the same I is assigned coefficient 0 and the coefficient of the first occurrence is incremented.
5. This coefficient is now reduced mod 2;

   a. If the coefficient is 0, the list is searched again for the largest I. If all coefficients become 0, J is a cycle, and the program exits.
   b. If the coefficient of the leading term I is 1, then WFILE is searched for I.

      b_1. If I is found untagged, then J tags I, and the process terminates.

      b_2. If I $\leftarrow$ K is found, then d(K) is calculated by 2, put in admissible form by 3, and appended to LIST; return to 4 and continue as before.

      b_3. If I is not present, then WFILE is searched for shorter and shorter tails of I until finally some tail of I is found that is tagged, say

$$(i_p, \ldots, i_{s+1}) \leftarrow (m_p, \ldots, m_s)$$

   Then let $K = (i_1, \ldots, i_{p-1}, m_p, \ldots, m_s)$ and d(K) is appended to LIST. LIST is again searched for the largest term, and the program continues as before.

5.1 Modifications to difftag

5.1.1 Truncation of WFILE

In computing the LIST from d(J), only those elements K of WFILE with filtration(K) $\leq$ filtration(J) + 1 and dimension(K) $\leq$ dimension(J) $-$ 1 will be needed. Accordingly, after J and WFILE have been read in by difftag, WFILE will be truncated by deleting all rows of filtration > filtration(J) + 1 or dimension > dimension(J) − 1.

5.1.2 Modifications to Handle a Long LIST

Sometimes the calculations above create a long LIST. For example, the calculations in the 40-stem may involve a LIST of more than 2000 lines. By the 50-stem, we have a LIST of more than 10,000 lines, and in the 59-stem, a LIST of more than 65,000 lines. The process of searching LIST repeatedly for the maximum term, and reducing mod 2, would be very (machine) time-consuming. Therefore we handle LIST as follows. Integers M and m are chosen (typically M = 2000 and m = 100). When the number of terms in LIST exceeds M, LIST is read out to a file called TEMP. TEMP is then sorted, and the largest m of these are read back into the program as a list called SHORT; the largest term remaining in TEMP will be called CUT. Next, difftag works on SHORT as in 4 and 5 above, except that

Each new row that is $\leq$ CUT is appended to TEMP.

Each new row that is > CUT is appended to SHORT.

If difftag finds the leading term I in SHORT untagged in WFILE, then J tags I as above. If all coefficients in SHORT become 0, then TEMP is again sorted, the largest m terms are read into SHORT, and the program proceeds as before.

5.1.3 Modifications to Take Account of Periodicity

Suppose that $I = (i_1, i_2, \ldots, i_s)$ is tagged by $J = (j_1, j_2, \ldots, j_{s-1})$. Let n be the least power of 2 greater than the difference of initials $j_1 - i_1$. Let $I^* = (i_1 + n, i_2, \ldots, i_s)$ and $J^* = (j_1 + n, j_2, \ldots, j_{s-1})$. The assertion is that if I^* is a cycle that is not tagged by some term less than J^*, then I^* will be tagged by J^*.

At present we cannot prove the full strength of this periodicity assertion, but we want to use it anyway. For this purpose, an integer flag is defined as follows. For each sequence $K = (k_1, k_2, \ldots k_s)$, let

$$\mu(K) = \begin{cases} k_1 - k_2 - 1 & \text{if } 2k_1 < k_2 \\ 0 & \text{otherwise} \end{cases}$$

Suppose that I is tagged by J, with $\Sigma_\alpha I_\alpha = d(\Sigma_\beta J_\beta)$. Then let flag($I \leftarrow J$) be the maximum of $\mu(K)$, where K appears in any relation that is used to express $d(\Sigma_\beta J_\beta)$ as a linear combination of admissable monomials. That is, flag($I \leftarrow J$) is the largest initial that is affected by the relations in the first position.

Lemma

Suppose that $I = (i_1, i_2, \ldots, i_s)$ is tagged by $J = (j_1, j_2, \ldots, j_{s-1})$, and suppose that flag($I \leftarrow J$) is less than i_1. Let $n = 2^k$ be the least power of 2 for which $2^k > j_1 - i_1$ and $I^* = (i_1 + n, i_2, \ldots, i_s)$ and $J^* = (j_1 + n, j_2, \ldots, j_{s-1})$; suppose also that I^* and J^* both appear in the table and that I^* is not tagged by some term earlier than J^*. Then J^* will tag I^*.

Proof. (This is proved in [8]; to make this chapter self-contained, we repeat the short proof here.) Let $M = H_*(RP^\infty)$, as a module with the Steenrod algebra acting on the right; as a vector space, M has a generator e(n) in every positive dimension. Consider the chain complex $M \otimes \Lambda$ with differential

$$d(x \otimes \lambda_I) = \textstyle\sum_j Sq^j \otimes \lambda_j \lambda_I + x \otimes d(\lambda_I)$$

For any sequence $I = (i_1, i_2, \ldots i_s)$, let PI stand for $e(i_1) \otimes (i_2, \ldots i_s)$. The map $M \otimes \Lambda \to \Lambda$ which sends PI to I is a map of chain complexes because of the formulas for $d(e_j)$ and $d(\lambda_j)$. For each positive integer m, let $M(m,\infty) = H_*(RP_m^{\ \infty})$. The assumption on flag($I \leftarrow J$) implies that $\Sigma_\alpha PI_\alpha = d(\Sigma_\beta PJ_\beta)$ in $M(i_1,\infty) \otimes \Lambda$, and James periodicity for truncated projective spaces implies that $\Sigma_\alpha PI_\alpha^* = d(\Sigma_\beta PJ_\beta^*)$ in $M(i_1 + n, \infty) \otimes \Lambda$. This shows that $\Sigma_\alpha I_\alpha^* = d(\Sigma_\beta J_\beta^*)$ in $\Lambda / \Lambda(i_1)$. Hence I^* will be tagged by J^* in Λ.

To check the validity of the periodicity assertion each time it occurs, the program difftag keeps track of this integer flag while it is calculating that I is tagged by J. If the flag is sufficiently small, $I^* \leftarrow J^*$ is placed in a file called STORE, for use at the proper time.

In the simplest version of the program (below), the program kill does not make use of this horizontal periodicity. A much faster version (also below) takes account of and stores the valid cases of horizontal periodicity, as checked by difftag. For this we use two more programs, postpone and pkill, which take account of the (validated) horizontal periodicities of period 2, 4, 8, 16, in increasing order. We have observed that in almost all cases, difftag calculates that flag($I \leftarrow J$) is less than i_1, which validates the periodicity. Furthermore (through stem 58), we have found no instances where the horizontal periodicity is incorrect.

6. THE TABLE FOR $E_2(S^3)$ TO STEM 58

Using these programs, we have calculated the unstable Adams E_2 terms for all spheres S^n and all stems through stem 58. In the pages that follow, we present the table for the sphere S^3 through stem 58. The symbol * stands for the sequence 2 4 1 1; a sequence of dots with 2's at each end stands for repeated 2's, where each dot substitutes for a missing 2. Certain subsequences occur so often that they are best written in compressed form: 6653 for 6 6 5 3; 24333 for 2 4 3 3 3; 45333 for 4 5 3 3 3; and 35733 for 3 5 7 3 3.

Table 1 $E_2^{s,t}(S^3)$

	1	2	3	4	5	6	7
4					2 1 1 1		
3			1 1 1	2 1 1			
2		1 1	2 1				
1	1						

	8	9	10	11	12	13
8						2 1 1 * 1
7				1 1 * 1	2 1 * 1	
6			1 * 1	2 * 1	2 1 1 2 3 3	
5		* 1	1 1 2 3 3	2 1 2 3 3		
4		1 2 3 3	2 2 3 3	2 3 3 3		
3	2 3 3					
2						
1						

	14	15	16	17	18
10					1 * * 1
9				* * 1	1 1 2 34411 1
8				1 2 34411 1	2 2 34411 1
7			2 34411 1	1 1 24333	2 1 24333
6			1 24333	2 24333	
5	2 3 3 3 3	24333			
4					
3					
2					
1					

	19	20	21	22
12			2 1 1 * * 1	
11	1 1 * * 1	2 1 * * 1		
10	2 * * 1	2 1 1 2 34411 1		
9	2 1 2 34411 1			
8	2 1 1 24333			
7				2 2 45333
6		2 45333		
5				
4				
3				
2				
1				

Table 1 (Continued)

14				1 * * * 1
13			* * * 1	1 1 2 34411 * 1
12			1 2 34411 * 1	2 2 34411 * 1
11		2 34411 * 1	1 1 * 24333	2 1 * 24333
10		1 * 24333	2 * 24333	
9	* 24333			2..2 45333
8		2.2 45333		
7			2 2 35733	2 3 35733
6	2 35733		1 23577	2 23577
5		23577	2 4 5 7 7	
4				
3				
	23	24	25	26

16			2 1 1 * * * 1	
15	1 1 * * * 1	2 1 * * * 1		
14	2 * * * 1	2 1 1 2 34411 * 1		
13	2 1 2 34411 * 1			
12	2 1 1 * 24333			
11				2....2 45333
10		2...2 45333		
9			2..2 35733	
8	2.2 35733		1 2 3 36653	2 2 3 36653
7	2 1 23577	2 3 36653		2 2 4 3577
6	2 3 3577	2 4 3577		
5				
4				
	27	28	29	30

18				1 * * * * 1
17			* * * * 1	1 1 2 34411 * * 1
16			1 2 34411 * * 1	2 2 34411 * * 1
15		2 34411 * * 1	1 1 * * 24333	2 1 * * 24333
14		1 * * 24333	2 * * 24333	
13	* * 24333			2......2 45333
12		2.....2 45333		
11			2....2 35733	
10	2...2 35733			2..2 3 36653
9		2.2 3 36653		
8	2 3 3 36653			2 1 2 35777
7	2 4 3 3577	1 2 35777	2 2 35777	
6	2 35777	2 45777		
5				
	31	32	33	34

Table 1 (Continued)

20			2 1 1 * * * * 1	
19	1 1 * * * * 1	2 1 * * * * 1		
18	2 * * * * 1	2 1 1 2 34411 * * 1		
17	2 1 2 34411 * * 1			
16	2 1 1 * * 24333			
15				2........2 45333
14		2.......2 45333		
13			2......2 35733	
12	2.....2 35733			2....2 3 36653
11		2...2 3 36653		
10			2 24333 6653	
9	24333 6653			2 3 5 5 36653
8				
7				2 3 5 7 7 7 7
6				
5				
	35	36	37	38

22				1 * * * * * 1
21			* * * * * 1	1 1 2 34411 * * * 1
20			1 2 34411 * * * 1	2 2 34411 * * * 1
19		2 34411 * * * 1	1 1 * * * 24333	2 1 * * * 24333
18		1 * * * 24333	2 * * * 24333	
17	* * * 24333			2..........2 45333
16		2.........2 45333		
15			2........2 35733	
14	2.......2 35733			2......2 3 36653
13		2.....2 3 36653		
12			2.2 24333 6653	
11	2.24333 6653		1 1 24733 6653	2 1 24733 6653 1 2 24733 6653
10		2 2 3 5 5 36653 1 24733 6653	2 24733 6653	
9	24733 6653			2 3 3 5 7 3577
8	2 3 5 7 3577	2 4 5 7 3577		
7	2 4 5 7 7 7 7			
6				
5				
	39	40	41	42

Table 1 (Continued)

24			2 1 1 * * * * * 1	
23	1 1 * * * * * 1	2 1 * * * * * 1		
22	2 * * * * * 1	2 1 1 2 34411 * * * 1		
21	2 1 2 34411 * * * 1			
20	2 1 1 * * * 24333			
19				2............2 45333
18		2...........2 45333		
17			2..........2 35733	
16	2.........2 35733			2........2 3 36653
15		2.......2 3 36653		
14			2....24333 6653	
13	2...24333 6653			2 2 1 2 24733 6653
12	2 1 1 24733 6653	2 1 2 24733 6653		
11			1 2 45553 6653	2 2 45553 6653
10	2 3 3 6 5 23577	2 45553 6653	1 2 3 3 5 93577	2 2 3 3 5 9 3577
9	2 4 3 5 7 3577	2 3 3 5 9 3577	2 4 3 5 9 3577	
8	2 3 3 59777	2 4 3 59777		
7				
6				
	43	44	45	46

26				1 * * * * * * 1
25			* * * * * * 1	1 1 2 34411 * * * * 1
24			1 2 34411 * * * * 1	2 2 34411 * * * * 1
23		2 34411 * * * * 1	1 1 * * * * 24333	2 1 * * * * 24333
22		1 * * * * 24333	2 * * * * 24333	
21	* * * * 24333			2..............2 45333
20		2.............2 45333		
19			2...........2 35733	
18	2...........2 35733			2..........2 3 36653
17		2.........2 3 36653		
16			2......24333 6653	
15	2.....24333 6653		1 1 * 24733 6653	2 1 * 24733 6653 1 2 * 24733 6653
14		2.2 1 2 24733 6653 1 * 24733 6653	2 * 24733 6653	
13	* 24733 6653		1 2.2 45553 6653	2..2 45553 6653
12	1 2 2 45553 6653	2.2 45553 6653		1 2 43565 23577
11		2 3 3 5 6 5 23577	2 43565 23577	
10			2 3 5 3 5 9 3577	2 4 5 3 5 9 3577 2 2 3 5 3 59777
9		2 3 5 3 59777	2 4 5 3 59777	
8				
7				
	47	48	49	50

Table 1 (Continued)

28			2 1 1 * * * * * * 1	
27	1 1 * * * * * * 1	2 1 * * * * * * 1		
26	2 * * * * * * 1	2 1 1 2 34411 * * * * 1		
25	2 1 2 34411 * * * * 1			
24	2 1 1 * * * * 24333			
23				2...............2 45333
22		2..............2 45333		
21			2.............2 35733	
20	2............2 35733			2...........2 3 36653
19		2..........2 3 36653		
18			2........24333 6653	
17	2.......24333 6653			2 2 1 2 * 24733 6653
16	2 1 1 * 24733 6653	2 1 2 * 24733 6653		
15			1 2...2 45553 6653	2....2 45553 6653
14	1 2..2 45553 6653	2...2 45553 6653		1 2.2 43565 23577
13		1 2 2 43565 23577	2.2 43565 23577	
12	2 2 43565 23577		2 3 35565 23577	2 4 35565 23577
11	2 2 3 5 3 5 9 3577			2 4 2 3 5 3 59777 2 3 4 6 3 5 9 3577
10				1 2 3 5 10 11 3577
9			2 3 5 10 11 3577	2 4 8 5 59777
8			2 4 6 9 11 7 7 7	
7				
6				2 4 7 11 15 15
5				
4				
3				
2				
1				
	51	52	53	54

Table 1 (Continued)

30				1 * * * * * * * 1
29			* * * * * * * 1	1 1 2 34411 * * * * * 1
28			1 2 34411 * * * * * 1	2 2 34411 * * * * * 1
27		2 34411 * * * * * 1	1 1 * * * * * 24333	2 1 * * * * * 24333
26		1 * * * * * 24333	2 * * * * * 24333	
25	* * * * * 24333			2..................2 45333
24		2.................2 45333		
23			2...............2 35733	
22	2...............2 35733			2.............2 3 36653
21		2.............2 3 36653		
20			2..........24333 6653	
19	2.........24333 6653		1 1 * * 24733 6653	2 1 * * 24733 6653 1 2 * * 24733 6653
18		2.2 1 2 * 24733 6653 1 * * 24733 6653	2 * * 24733 6653	
17	* * 24733 6653		1 2.....2 45553 6653	2......2 45553 6653
16	1 2....2 45553 6653	2.....2 45553 6653		1 2...2 43565 23577
15		1 2..2 43565 23577	2...2 43565 23577	
14	2..2 43565 23577		1 2 2 4 35565 23577	2.2 4 35565 23577 2 1 1 2 42353 59777
13	1 2 4 35565 23577	2 2 4 35565 23577 1 1 2 42353 59777		
12	1 2 42353 59777	2 2 42353 59777	2 1 1 2 3 5 10 11 3577	
11	2 4 5 5 3 5 9 3577 1 1 2 3 5 10 11 3577	2 1 2 3 5 10 11 3577 1 1 2 4 8 5 59777	2 1 2 4 8 5 59777	
10	2 2 3 5 10 11 3577 1 2 4 8 5 59777	2 3 3 5 10 11 3577 2 2 4 8 5 59777	2 3 4 8 5 59777 2.2 4 6 9 11 7 7 7	2 4 5 9 3 59777
9	2 2 4 6 9 11 7 7 7	1 2 3 5 9 15 7 7 7	2 2 3 5 9 15 7 7 7	2 3 3 5 9 15 7 7 7 2 1 1 2 4 7 11 15 15
8	2 3 5 9 15 7 7 7	2 4 5 9 15 7 7 7 1 1 2 4 7 11 15 15	2 3 5 7 11 15 7 7 2 1 2 4 7 11 15 15	2 4 5 7 11 15 7 7
7	1 2 4 7 11 15 15	2 2 4 7 11 15 15	2 3 4 7 11 15 15	2 3 5 7 11 15 15
6				
5				
4				
3				
2				
1				
	55	56	57	58

REFERENCES

1. J. F. Adams, On the structure and applications of the Steenrod algebra, Comment. Math. Helv., 52, (1958), 180–214.
2. M. G. Barratt, EHP calculations of homotopy groups of spheres (unpublished tables), (1960).
3. A. K. Bousfield and E. B. Curtis, A spectral sequence for the homotopy of nice spaces, Trans. Am. Math Soc., 151, (1970), 457–479.
4. A. K. Bousfield, E. B. Curtis, D. M. Kan, D. G. Quillen, D. L. Rector, and J. W. Schlesinger, The mod-p-lower central series and the Adams spectral sequence, Topology, 5, (1966), 331–342.
5. E. B. Curtis, Lectures on Simplicial Homotopy, Matematisk Institut, Aarhus, Denmark (1968).
6. E. B. Curtis, Some non-zero homotopy groups of spheres, Bull. Am. Math. Soc., 75, (1969), 541–546.
7. E. B. Curtis, Simplicial homotopy theory, Adv. Math., 6, (1971), 107–209.
8. E. B. Curtis, P. Goerss, M. Mahowald, and R. J. Milgram, Calculations of unstable Adams E_2 terms for spheres, Proc. of the Seattle Conference in Algebraic Topology, (1985), Lecture Notes in Math., 1286, Springer-Verlag, New York (1987), 208–266.
9. M. E. Mahowald and M. C. Tangora, Some differentials in the Adams spectral sequence, Topology, 6, (1967), 349–369.
10. J. P. May, The cohomology of the Steenrod algebra; stable homotopy groups of spheres, Bull. Am. Math. Soc., 71, (1967), 377–380.
11. M. C. Tangora, On the cohomology of the Steenrod algebra, Math. Z., 116, (1970), 18–64.
12. M. C. Tangora, Some remarks on the lambda algebra, Proc. Evanston Conference on Algebraic Topology (1977), Lecture Notes in Math., 658, Springer-Verlag, New York (1978), 476–487.
13. M. C. Tangora, Generating Curtis tables, Proc. Canadian Summer Congress on Algebraic Topology, Vancouver B. C. (1977), Lecture Notes in Math., 673, Springer-Verlag, New York (1978), 243–253.

14. M. C. Tangora, Computing the homology of the lambda algebra, Mem. Am. Math. Soc., 337 (1985).

15. H. Toda, Composition methods in homotopy groups of spheres, Annals of Math. Studies, No. 49, Princeton University Press (1962).

16. J. S. P. Wang, On the cohomology of the mod-2 Steenrod algebra, Illinois J. Math., 11, (1967), 480–490.

17. R. J. Wellington, The unstable Adams spectral sequence for free iterated loop spaces, Mem. Am. Math. Soc., 258 (1982).

18. R. J. Wellington, The computation of Ext groups for modules over the Steenrod algebra, Preprint (1982).

19. G. W. Whitehead, (absolutely amazing) unpublished tables (c. 1970).

6

Use of a Computer to Suggest Key Steps in the Proof of a Theorem in Topology

DONALD M. DAVIS

Lehigh University
Bethlehem, Pennsylvania

1. IMMERSIONS OF PROJECTIVE SPACES

Real projective n-space P^n is the space obtained from the n-sphere S^n by identifying antipodal points $(x \sim -x)$. Although S^n is naturally embeddable in Euclidean space R^{n+1}, the identification prevents P^n from being embeddable in R^{n+1}, and a natural question is: what is the smallest Euclidean space in which P^n can be embedded? A closely related question, which lends itself more readily to methods of algebraic topology, is to find the smallest Euclidean space in which P^n can be immersed.

A (differentiable) n-manifold is a topological space which is locally homeomorphic to R^n in such a way that the homeomorphisms that overlap do so in a differentiable way. S^n and P^n are n-manifolds. An immersion of an n-manifold M in R^k is a function $M \to R^k$ which, when viewed locally (using the homeomorphisms) as maps $R^n \to R^k$, has rank-n Jacobian matrices at all points. The familiar picture of the Klein bottle K is of an immersion in R^3; K cannot be embedded in R^3.

Hassler Whitney made these notions accessible around 1940 and proved that every n-manifold can be immersed in R^{2n-1} [14]. Ralph Cohen [4] recently strengthened this result by proving that every n-manifold can be immersed in $R^{2n-\alpha(n)}$, where $\alpha(n)$ denotes the number of 1's in the binary expansion of n. This was known to be best possible, for a Cartesian product of P^{2^e}'s corresponding to the binary expansion of n is an n-manifold which does not immerse in $R^{2n-\alpha(n)-1}$. Cohen's result gave no new information for projective spaces because Milgram had proved in 1967 [12] that P^n immerses in $R^{2n-\alpha(n)}$; for many values of n, P^n immerses in much smaller Euclidean spaces.

A survey of known results and methods for studying immersions and embeddings of projective spaces was provided by James in 1971 [10], and an updated tabulation of results appeared in [5]. The first unknown question is whether P^{24} immerses in R^{38}. Virtually all previous immersion and nonimmersion results have applied only to restricted values of $\alpha(n)$ and restricted congruences (usually mod 8) of n. However, I recently proved the following strong result [6].

Theorem 1.1

P^{2n} cannot be immersed in $R^{4n-4b-2\alpha(n-b)}$, where b is the smallest nonnegative integer such that $\alpha(n - b) \leq b + 1$.

For example, n = 15 and b = 2 imply that P^{30} cannot be immersed in R^{46}, a strong result first proved in [1]. The importance of Theorem 1.1 is that it applies across the board to give good results for all n. It is in a certain sense within two dimensions of all known nonimmersion results and improves on known results by arbitrarily large amounts in certain cases. Theorem 1.1 follows readily from the more elegant result that $P^{2(m+\alpha(m)-1)}$ cannot be immersed in $R^{4m-2\alpha(m)}$. Although the proof of Theorem 1.1 does not use any computer calculations, the computer was instrumental in suggesting several key steps in the proof.

2. REDUCTION FROM TOPOLOGY TO ALGEBRA

The reduction from a problem in topology to one in algebra is standard, following most closely that of [10, section 6] and [1]. Using vector bundles as an intermediate tool, one can show that if P^{2n} immerses in R^{2k}, then there exists an axial map

$$f : P^{2^L-2-2k} \times P^{2n} \to P^{2^L-2n-1} \qquad [2.1]$$

where L is a sufficiently large integer (e.g., L > k). An <u>axial map</u> is one which, when restricted to each factor, is homotopic to the identity map.

Heinz Hopf [9] introduced axial maps and used ordinary cohomology $H^*(\ )$ to prove a nonexistence theorem for axial maps. A continuous map $f: Y \to P$ induces a ring homomorphism $f^*: H^*(P) \to H^*(Y)$, where $H^*(Y)$ is the graded ring $\oplus H^i(Y)$. $H^*(P^{2m})$ is a truncated polynomial algebra $Z_2(X)/(X^{m+1})$ on a generator X of degree 2, i.e.,

$$H^i(P^{2m}) = \begin{cases} Z_2 = Z/2 & \text{if } i = 2j \text{ with } 0 \leq j \leq m \\ 0 & \text{otherwise} \end{cases}$$

with nonzero classes X^j. $H^*(P^{2^L-2-2k} \times P^{2n})$ has classes $X_1{}^iX_2{}^j$ of order 2 for $i \leq 2^{L-1} - 1 - k$ and $j \leq n$, while $X^{2^{L-1}-n}$ is 0 in $H^*(P^{2^L-2n-1})$. If the axial map f of (2.1) exists, it must satisfy $f^*(X) = X_1 + X_2$ by the axial property, and hence

$$0 = f^*(X^{2^{L-1}-n}) = (X_1 + X_2)^{2^{L-1}-n} = \sum_{j=k+1-n}^{n} \binom{2^{L-1}-n}{j} X_1{}^{2^{L-1}-n-j}X_2{}^j$$

Thus if for some j between $k + 1 - n$ and n, the binomial coefficient $\binom{2^{L-1} - n}{j}$ is odd, we may deduce that P^{2n} cannot be immersed in R^{2k}. It is well known that $\nu\binom{2^{L-1} - n}{j} = \alpha(j) + \alpha(n - 1) - \alpha(n + j - 1)$, where $\nu(\ )$ denotes the exponent of 2. This implies Hopf's nonimmersion of P^p in R^{2p-2} if p is a 2-power.

We mimic Hopf's proof, except that we use a cohomology theory $B^*(\ )$, which contains more information than $H^*(\ )$. $B^*(\ )$ is the (generalized) cohomology theory associated to the spectrum BP<2> [2,11], which is derived from complex cobordism theory. All we need to know here is the $B^*(\ )$-cohomology groups. Let $Z_{(2)}$ denote the integers localized at 2, i.e., the rational numbers with odd denominators, and V the polynomial algebra over $Z_{(2)}$ on a generator v of degree -6. Then in even degrees $B^*(P^{2n-1})$ is an algebra over V generated by an element X of degree 2 with relations $2^{n-i}X^i$, $1 \leq i < n$. There is a power series $A(X) = \sum_{i \geq 0} \alpha_i X^{3i-1}$ with coefficients α_i in the ring $\hat{Z_2}$ of 2-adic integers so that in even degrees greater than 2k and 2m

$$B^*(P^{2k} \times P^{2m}) = \frac{V[X_1,X_2]}{(X_1{}^{k+1},X_2{}^{m+1},A(X_1) - A(X_2))} \qquad [2.2]$$

where X_1 and X_2 have degree 2. The series $A(X)$ is closely related to the 2-series of formal group theory; it is effectively calculable, but many minutes of DEC-20 computer time were required for Pascal and REDUCE programs to calculate $\alpha_i \bmod 2^{22-3i}$ for $i \leq 7$, the limit of my specific knowledge of $A(X)$.

Let $V_{k,m}$ denote the graded ring of (2.2) and $V^i_{k,m}$ its component in degree 2i. As in Hopf's application, if one can show that $(X_1 + X_2)^{2^{L-1}-n}$ is nonzero in $V_{2^{L-1}-1-k,n}$, then one can deduce that P^{2n} does not immerse in R^{2k}. Theorem 1.1 follows by straightforward combinatorics from the following result, the proof of which is discussed in the next section.

Theorem 2.3

If $\nu\binom{n+s}{k-s} = s$, then $(X_1 + X_2)^n$ is nonzero in $V^n_{k,n-k+3s}$

3. USE OF THE COMPUTER IN SOLVING THE ALGEBRAIC PROBLEM

The group $V^n_{k,n-k+3s}$ is independent of n and k. Denote it by G_s. Prior approaches to the problem [1,3] had suggested the plausibility of Theorem 2.3. My published proof of it [6] describes very explicitly the structure of the abelian group G_s, but my first proof consisted of proving

Proposition 3.1

There exists a homomorphism $g_s : G_s \to Z/2^{s+1}$ such that $g_s(X_1{}^{k-j}X_2{}^{n-k+j}) = \left[\begin{matrix} s \\ j-s \end{matrix}\right]$ for $0 \leq j \leq 3s$.

A standard combinatorial identity then implies Theorem 2.3.

G_s can be presented by a matrix M_s whose columns correspond to $v^iX_1{}^{k-j}X_2{}^{n-k+j+3i}$, in which the (i,j)'s are ordered as (s,0), (s − 1,0), (s − 1,1), (s − 1,2), (s − 1,3), (s − 2,0),...,(s − 2,6), (s − 3,0), ...,(1,3s − 3), (0,0),...,(0,3s). Thus, for example, the third column corresponds to $v^{s-1}X_1{}^{k-1}X_2{}^{n-k+1+3(s-1)}$. The rows present the coefficients in relations $R_{i,j}$, $0 \leq i \leq s$, $0 \leq j \leq 3(s-1)$, ordered in the same manner as the columns, where $R_{i,j}$ is $(A(X_1) - A(X_2))v^iX_1{}^{k-j}X_2{}^{n-k+j+3i+1}$.

The matrices M_s have a nice form, consisting of $(3i + 1) \times (3j + 1)$ submatrices $M_{i,j}$, whose only nonzero elements are $-\alpha_{i-j}$ on the main diagonal (as far as it goes) and α_{i-j} $3(i - j) - 1$ spaces below. However, this form is not convenient for showing that there is a vector

$$U = (u_{s,0}, u_{s-1,0}, \ldots, u_{1,3s-3}, 0, \ldots, 0 \binom{s}{0}, \ldots, \binom{s}{s}, 0, \ldots, 0)$$

whose inner product with each row of M_s is 0 mod 2^{s+1}, which is equivalent to Proposition 3.1. From the facts that $\alpha_0 = 2$ and α_1 is odd, and the form of M_s, it is easy to see that 3.1 requires knowledge of α_i mod 2^{s+1-i}.

Early attempts to prove Proposition 3.1 consisted of row-reducing M_s by hand for $s \leq 4$, sometimes using specific values of α_i's and sometimes using literal values to try to see patterns for a proof. At least for $s \leq 3$ I could see that vectors U existed as desired, but by $s = 4$ my work was very error-prone. It was natural to let a computer perform the row reductions, which was done quite easily for $s \leq 7$, in which range the coefficients α_i were known to the required 2-power accuracy. For $s = 8$, there were four possible series A(X); I didn't know α_8 mod 2, and I knew α_7 mod 2 but needed it mod 4. The computer discovered that a solution U exists if α_8 is even but not if α_8 is odd. Other computer experiments using modifications of A(X) suggested that solutions come much more easily if α_{2i} is always even. This was the first key step suggested by the computer—that the situation would be simpler if α_{2i}'s are always even.

The second key step suggested by the computer was to replace the submatrices $M_{i,j}$ by polynomials. Of course, the computer didn't print out "USE POLYNOMIALS"; what it did was to show me that if the row reductions are performed in a systematic manner, then the submatrices will be Toeplitz ($m_{i+k,j+k} = m_{i,j}$ for all i,j,k) at certain stages of the row reduction. Thus a submatrix is determined by the polynomial whose coefficients are the entries on its top row and those in its left column.

The Toeplitz property was certainly not clear from my haphazard hand row reductions, and since the computer usually reduced different columns modulo different 2-powers, it wasn't totally clear from most computer printouts, but on one occasion I asked it to print certain stages of the row reduction of M_5, and here the Toeplitz property became quite discernible.

The use of polynomials was crucial in both proofs of Theorem 2.3. The first proof required solving the matrix equation

$$\begin{bmatrix} p_1 & p_0 & 0 & \cdots 0 & \\ p_2 & p_1 & p_0 & 0 \cdots 0 & \\ & & & & \\ p_{s-1} & \cdots & p_1 & p_0 & \\ p_s & \cdots & p_2 & p_1 & \end{bmatrix} \cdot \begin{bmatrix} \varphi_0 \\ \varphi_1 \\ \cdot \\ \cdot \\ \varphi_{s-1} \end{bmatrix} = \begin{bmatrix} 0 \\ \cdot \\ \cdot \\ 0 \\ z^s(1+z)^s p_0 \end{bmatrix}$$

where $p_i = \alpha_i(z^{3i-1} - 1)$ and solutions $\varphi_i(z) = \sum u_{s-i,j} z^j$ must have $\deg(\varphi_i) \le 3i$. Cramer's rule showed that this was possible as long as α_{2i}'s are even. I then asked some experts if they knew whether α_{2i} might always be even. Steve Mitchell pointed out how to use a formal group formula [15,3.17] with which I was familiar to prove an analogous statement, and it then took only 2 days to complete the proof.

4. SOME MORE COMPUTER DETAILS

The computer work in this project consisted of two parts: (i) calculation of the series A(X) in (2.2) and (ii) row reduction of matrices discussed in Section 3. The latter was more crucial to the project but probably less interesting to the computer person. For this, Pascal, the language of choice at Lehigh, was used on a mainframe DEC-20 computer. The matrices being row-reduced were large (84×92 was a common size), but they had lots of 0's, and their columns were considered mod various small 2-powers.

Calculating A(X) was a two-step process. First the 2-series t(X), which is a series in $Z[v_1, v_2][[X]]$, was calculated, and then t(X) was used to determine a "reduced" 2-series, in which v_1 is not present.

The calculation of t(X) is standard (e.g., [15]). It is determined as a series in $Z[m_1, m_2, \ldots][[X]]$ by the equations $t(X) = \exp(2 \cdot \log X)$, $\exp(\log X) = X$, and $\log X = X + \sum m_i X^{2^i}$. The m_i's are then written in terms of v_i's by

$$2m_n = v_n + \sum_{i=1}^{n-1} m_i v_{n-i}^{2^i}$$

To get the series involving just v_1 and v_2, we set $v_i = 0$ for $i > 2$.

These sorts of manipulations lend themselves to the program REDUCE [8], to which I had been introduced by a member of Lehigh's math department, Jerry Rayna. I have found REDUCE to be excellent for polynomial manipulations and working with fractions. It is quite similar to MACSYMA, which I had used at the Institute for Advanced Study (see [13]). Although I didn't need fractions here, earlier work [3] used similar calculations with a different set of generators and required fractions.

Extensive calculations of t(X) had been published in [7], but I needed to know more. The REDUCE program which performs them is short and self-explanatory ($\ell r = \log(X)/X$, $\ell p(i) = \ell r^i$).

```
m(0) := 1
v(0) := 2
for n := 1:4 do m(n) := (for i := 0:(n - 1) sum(m(i) * v(n - i)^
    (2^i)))/2
let x^22 = 0
ℓr := (for j := 0:4 sum(m(j) * x^(2^j - 1)))
ℓp(0) := 1
for i := 1:21 do ℓp(i) := ℓp(i - 1) * ℓr
e(0) := 1
for i := 1:21 do << coeff((for j := 0:(i - 1)sum(e(j) * ℓp(j + 1) *
    x^j)) , x,c); e(i) := -c(i) >>
write t := (for j := 0:21 sum(e(j) * (2 * x)^(j + 1) * ℓp(j + 1))
```

This is elegant, but it is also slow and uses lots of memory. By itself, REDUCE was unable to calculate much beyond [7]. I ended up writing a Pascal program to calculate t(X) in terms of the m's and then using REDUCE to rewrite it in terms of v's.

A(X) is related to t(X) by a result I proved in [6]: For any d, there is a series u(X) with coefficients in v_1 and v_2 and leading coefficient 1 such that

$$u(X) \cdot t(X) = av_1X^2 + \sum_{i \geq 0} \alpha_i v_2{}^i X^{3i+1} \mod (2^{d-m}X^m) \qquad [4.1]$$

with a odd and $\alpha_0 = 2$. Then $A(X) = \sum \alpha_i v_2^i X^{3i-1}$. The requirement in (4.1) is that coefficients of $v_1^i v_2^j X^{i+3j+1}$ are 0 if $ij > 0$ or $i > 1$. This results in a system of linear equations for the coefficients of u(X), which were solved by a Pascal program performing row reduction.

REFERENCES

1. L. Astey, Geometric dimension of bundles over real projective spaces, Q. J. Math. Oxford, 31, (1980), 139—155.
2. N. A. Baas, On bordism theory of manifolds with singularities, Math. Scand., 33, (1973), 279—302.
3. M. Bendersky and D. M. Davis, Unstable BP-homology and desuspensions, Am. J. Math., 107, (1985), 833—852.
4. R. Cohen, Immersions of manifolds, Ann. Math., 122, (1985), 237—328.
5. D. M. Davis, Some new immersions and nonimmersions of real projective spaces, Proc. Northwestern Homotopy Theory Conf., Contemporary Math, 19, American Mathematical Society, Providence, R.I. (1982), 51—64.
6. D. M. Davis, A strong nonimmersion theorem for real projective spaces, Ann. Math., 120, (1984), 517—528.
7. V. Giambalvo, Some tables for formal groups and BP, Lecture Notes in Math., 658, Springer-Verlag, New York (1978), 169—176.
8. A. C. Hearn, REDUCE User's Manual, Version 3.2, RAND Corp., Santa Monica, Calif. (1985).
9. H. Hopf, Ein topologischer Beitrag zur reelen 'Algebra', Comment. Math. Helv., 13, (1941), 219—239.
10. I. M. James, Euclidean models of projective spaces, Bull. London Math. Soc., 3, (1971), 257—276.
11. D. C. Johnson and W. S. Wilson, Projective dimension and Brown-Peterson homology, Topology, 12, (1973), 327—353.
12. R. J. Milgram, Immersing projective spaces, Ann. Math., 85, (1967), 473—482.

13. N. J. A. Sloane, My Friend MACSYMA, Notices AMS, 33, (1986), 40–43.
14. H. Whitney, The singularities of a smooth n-manifold in (2n − 1)-space, Ann. Math., 45, (1944), 247–293.
15. W. S. Wilson, A BP introduction and sampler, CBMS Regional Conf. Series, 48 (1982).

7

Local Symmetry in the Plane
Experiment and Theory

PETER J. GIBLIN

University of Massachusetts at Amherst
Amherst, Massachusetts

University of Liverpool
Liverpool, England

1. INTRODUCTION

The computer, with a good graphical facility, is particularly well suited to the study of plane curves. But what new results of interest can there possibly be in that well-worked area? By a happy coincidence, an idea which arose in theoretical biology and pattern recognition (namely the idea of measuring local reflexional symmetry of a plane curve) leads to interesting problems in the theory of singular points of smooth functions ("singularity theory") which are just within the reach of present techniques to solve satisfactorily. See Section 3. Even more happily, some of the answers were suggested in advance by an experimental approach using computer graphics, and there is scope for further experiment, both with modest graphical facilities and unsophisticated programming (see especially the piecewise linear case in Section 2.4) and with more ambitious intentions. Of course, the corresponding problems for surfaces in three-dimensional space are even more challenging; on the theoretical side the "static" problem of studying a single surface is soluble [8] but there are serious obstacles to the study of families of surfaces. I have not yet applied computer graphics to the surface case.

The basic idea is to study a two-dimensional shape, bounded, let us say, by a smooth simple closed curve γ, by means of its local sym-

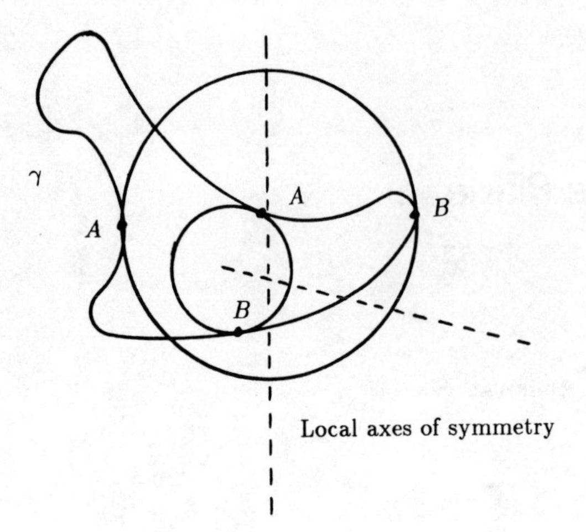

Figure 1 Bitangent circles and local symmetry axes.

metries about lines. Thus a line is a local (or "infinitesimal" or "first order") axis of symmetry if the tangents at two points A,B of the curve are symmetrical about the line (Figure 1). An equivalent statement is that a circle exists which is tangent to γ at A and at B; we call this a bitangent circle for γ.

We can capture all these local symmetries by drawing all the axes, or by following the path of the centers of the bitangent circles, which we call the symmetry set of γ. In fact, it is not hard to show that, roughly speaking, the envelope of axes of symmetry is the symmetry set. (The roughness of speech arises from the need to parametrize the family of axes before speaking of an envelope; at any rate the axes are all tangent to the symmetry set, interpreting this as a limiting tangency at a singular point of the symmetry set.)

The kind of symmetry measured by bitangent circles is a local reflectional symmetry; it was studied originally by the theoretical biologist H. Blum [3] and taken up under the name symmetric axis transform by M. Brady [5] for pattern recognition. See also [4] for another treatment. There is usually the condition that both contacts of the circle with the curve should be internal, or that the whole circle should be interior to the curve. Note, however, that here we consider all bitangent circles.

Of course, we could also study translational or rotational symmetry. For the former perhaps the locus of midpoints of chords between points

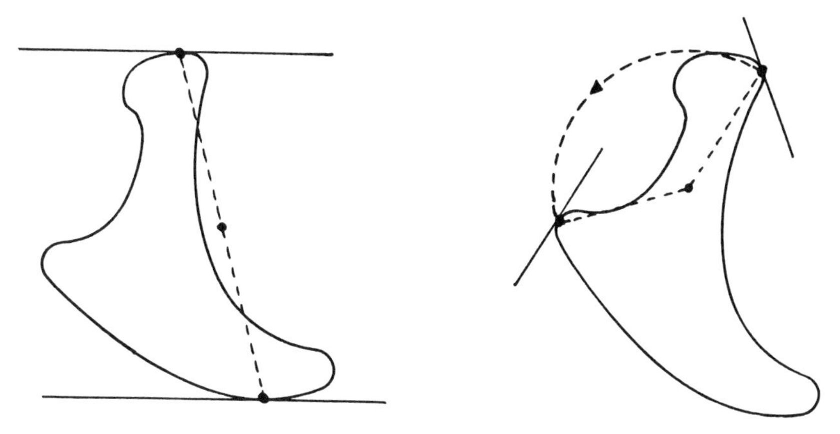

Figure 2 Measuring translational and rotational symmetry.

of contact of parallel tangents is the appropriate object to study. (See Figure 2, left, and [10, p. 703].) For the latter, we should look for centers c which rotate a point, its tangent line, and the circle of curvature from one position on the curve to the other. This amounts to rotating point and tangent line between points of equal curvature and sense (Figure 2, right). But here we shall concentrate on reflectional symmetry.

This chapter gives a nontechnical survey of research which is, or will be, reported in full detail elsewhere [1,2,8,9,10]; in addition, more is said here (see Section 2) about the computational techniques involved. In Section 3 we give an account of the mathematical underpinning of this work, and in Section 4 we sketch a global result about symmetry sets.

2. COMPUTATION

2.1 Finding Centers of Bitangent Circles

The most straightforward, though not necessarily the most efficient, way to find centers of bitangent circles is as follows (see Figure 3).

Let $\gamma : I \to R^2$ be a smooth parametrization of a simple curve in the plane (so $\gamma'(t)$ is never $0 \in R^2$), where I is an interval of R, or a circle for closed curves, and let t_1, t_2 be distinct points of I. Writing $\gamma(t) = (X(t), Y(t))$, the unit tangent $T = T(t)$ and the unit normal $N = N(t)$ are

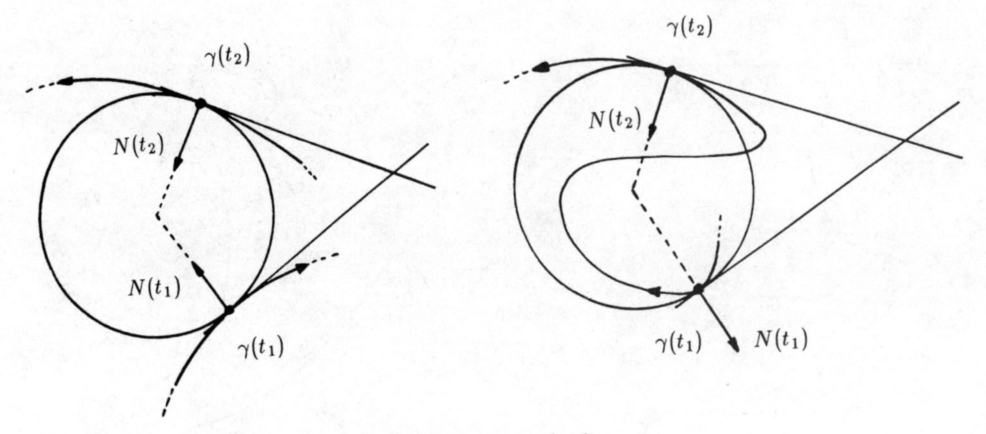

Figure 3 Finding centers of bitangent circles.

$$T = \frac{(X',Y')}{\sqrt{(X'^2 + Y'^2)}}$$

and

$$N = \frac{(-Y',X')}{\sqrt{(X'^2 + Y'^2)}}$$

The condition for a circle of radius r to be bitangent to γ at $\gamma(t_1)$ and $\gamma(t_2)$ is then

$$\gamma(t_1) \pm rN(t_1) = \gamma(t_2) + rN(t_2) \qquad \text{(= center of circle)} \qquad [1]$$

We can conveniently allow for the other possible pairs of signs by simply permitting r to be <0. Then the radius is $|r|$. The way to detect pairs t_1,t_2 for which an r (>0 or <0) exists is to rewrite (1) as

$$\gamma(t_1) - \gamma(t_2) = r(N(t_2) \mp N(t_1)) \qquad [2]$$

and to use the fact that $N(t_2) \mp N(t_1)$ is always perpendicular to $T(t_2) \mp T(t_1)$. Thus we search for solutions of

$$(\gamma(t_1) - \gamma(t_2)) \cdot (T(t_2) \mp T(t_1)) = 0 \qquad [3]$$

In Figure 3, − and + correspond respectively to the left and right pictures; we could call these the <u>negative case</u> (orientations coherent round the circle) and the <u>positive case</u> (orientations not coherent).

Conversely, if (3) with the upper sign (resp. lower sign) holds, and if $T(t_2) \neq T(t_1)$ (resp. $T(t_2) \neq -T(t_1)$), then we can deduce that (1) holds, for some $r \neq 0$. These exceptional cases do need to be excluded from the computations, for they make r indeterminate from (1). Also, solutions of (3) are found by the method of detecting sign changes in the left side: fix t_1 and let t_2 run in small steps through I; whenever the left side changes sign a solution lies between two consecutive values of t_2. (With a reasonably large number of subdivisions of I we can take one of these values, say the smaller, as the solution t_2 of (3); or of course some kind of interpolation method can be used to determine the solution more accurately than that.) There is a slight danger that the left side will <u>not</u> change sign at a zero: the left side has a multiple zero (when $T(t_2) \mp T(t_1) \neq 0$) if and only if the circle is <u>osculating</u> at $\gamma(t_2)$. (See [10], Figures 6 and 7, for an explanation.) These solutions will not show up, but since in fact the symmetry set has a <u>cusp</u> at such a point (see Section 3) other points of the symmetry set will accumulate near this missing center and no one will notice the gap.

2.2 Procedure for Plotting the Symmetry Set Point by Point

For each fixed $t_1 \in I$

1. Locate a solution $t_2 \neq t_1$ of (3) and reject it if $T(t_2) \mp T(t_1) = 0$, or perhaps if $|T(t_2) \mp T(t_1)|$ falls below some threshold value such as 0.001.
2. Calculate r from (2). Since r may be <0 it is best not to take absolute values but to use either of the two scalar equations corresponding to splitting into x or y coordinates. The one with larger denominator could be used for preference.
3. Find the required center as $\gamma(t_2) + rN(t_2)$ from (1), and put a dot there.

Naturally, every center is found twice by this procedure (corresponding to t_2, t_1 and to t_1, t_2), unless some device such as considering only $t_1 < t_2$ is adopted. More sophistication is needed to plot the symmetry set in continuous pieces; this is because the solution t_2 in 1 will

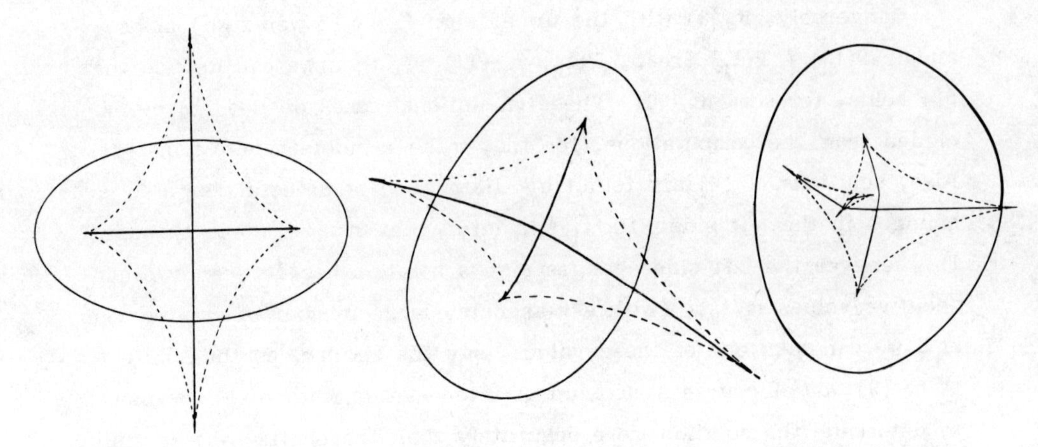

Figure 4 Symmetry sets (full line) and evolutes (dashed line) of an ellipse, and the curves $y^2 - 2bxy = a^2(x - x^3)$ for a = 1, b = 0.45 and a = 1.025, b = 0.09.

cease to be a function of t_1 when the circle becomes osculating at $\gamma(t_2)$. See [10], p. 694.

Figure 4 was obtained by this method, by Stephen Brassett.

2.3 Examples

After the ellipse, the simplest closed examples are cubic ovals, for instance, the curves

$$(y - bx)^2 = a^2(x - x^3) \qquad (0 \leq x \leq 1)$$

for constants a and b. This has a convenient parametrization $\gamma(t) = (X(t),Y(t))$ where $X(t) = \frac{1}{2} + \frac{1}{2}\cos t$, $Y(t) = \frac{1}{4}a \sin t \sqrt{(6 + 2\cos t)} + \frac{1}{2}b(1 + \cos t)$. Two of the examples of Figure 4 come from the closely related family $y^2 - 2bxy = a^2(x - x^3)$.

Many of the phenomena of symmetry sets can be illustrated by this family; however, some require nonconvex curves. If we drop the requirement of closure then

$$y = x^4 + ax^2 + bx$$

(parametrized by x) is a good family. Otherwise "distorted limaçons" of the form

$$x = a \cos t + \cos 2t + b \cos 3t$$

$$y = a \sin t + \sin 2t + c \sin 3t$$

are good examples.

2.4 The Piecewise-Linear Case

It is worth mentioning here that the corresponding ideas for "piecewise-linear curves," that is, simple polygons, are also very interesting and make excellent use of unsophisticated computer graphics. (See also Section 3; details of this joint work with T. F. Banchoff will appear elsewhere.) It is necessary to have a good definition for tangency between a circle and a polygon at a node (that is, vertex, but this word has undesirable overtones here) of the polygon. See Figure 5, noting that if a circle is tangent to an edge at a node and if the adjacent edge enters the circle, then the circle is deemed to have higher contact, that is, to be osculating. (Generically, a polygon will not have two edges in a straight line.) Note that the distance from the center of the circle to the polygon does have a local maximum or minimum at a point of ordinary tangency.

The locus of centers of bitangent circles is then a union of straight line segments and parabolic arcs, and a procedure for plotting these can be devised which starts with the circles bitangent to a particular pair (edge, edge), (edge, node), or (node, node) and traces the centers in an unbroken sequence of segments and parabolic arcs until the symmetry set closes up, homes in on a node, or goes out to infinity (Figure 6).

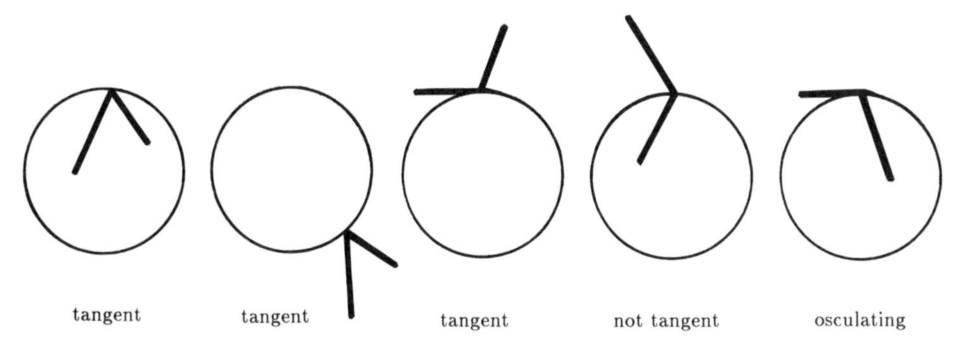

Figure 5 Tangency between a circle and a polygon.

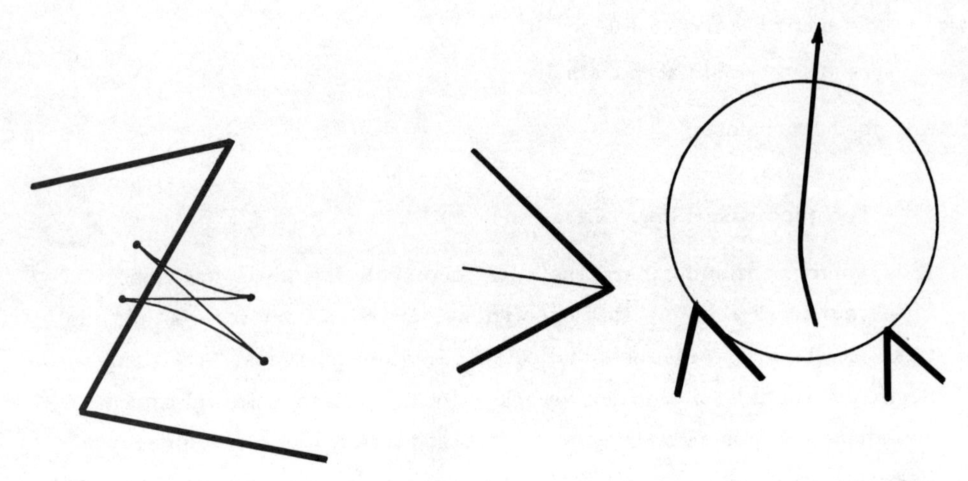

Figure 6 Symmetry set of a polygon closing up or ending or going off to infinity.

There are more efficient ways of plotting a closely related set, the "medial axis transform" (or "central set"), where the center x of a bi-tangent circle is included if and only if the circle is wholly contained in the compact region bounded by the polygon. See [12] for connections with Voronoi diagrams of polygons.

3. MATHEMATICAL TECHNIQUES AND GENERAL RESULTS ON SYMMETRY SETS

3.1 Bifurcation Sets

A little experiment with some relatively uncomplicated curves, such as those in Section 2.3 above, will soon produce the conviction that, locally, the symmetry set has only a small number of different forms. It is also very instructive to see the relationship between the symmetry set and the <u>evolute</u> (or <u>focal set</u>) of the curve γ, that is, the locus of centers of osculating circles (circles with highest contact at the points of the curve). This set is drawn dashed in Figure 4. If $\gamma(t) = (X(t), Y(t)$ then the center of the osculating circle at $\gamma(t)$ is $\gamma(t) + \rho(t)N(t)$ where ρ, the radius of curvature, is $(X'^2 + Y'^2)^{3/2}/(X'Y'' - X''Y')$. In fact, when the points $\gamma(t_1)$ and $\gamma(t_2)$ of contact between circle and curve tend to coincidence (t_1 and t_2 both tending to the same t_0) the limit circle has at least $2 + 2 = 4$ coincident points of contact with γ,

Figure 7 From left to right: Every point is the center of a bitangent circle (A_1^2); the dot is the center of an osculating circle tangent elsewhere (A_1A_2); the dot is the center of an osculating circle at a vertex (A_3); the dot is the center of a tritangent circle (A_1^3).

and its center is a vertex of γ where the radius of curvature ρ is stationary. The vertex shows up as a cusp on the evolute, so that the symmetry set comes to an end in a cusp of the evolute. In fact, local forms of the symmetry set (full line) and evolute (dashed line) are as in Figure 7. See below for the notation.

Some information of a fairly elementary kind related to these pictures is contained in [10]. But the mathematical technique for handling them in detail depends on regarding the symmetry set and evolute together as making up the full bifurcation set of a family of functions. To be specific, let

$$F : I \times R^2 \to R$$

be defined by $F(t,x) = ||\gamma(t) - x||^2$, the squared-distance function from $x \in R^2$ to $\gamma(t)$ on the curve. We regard F as a two-parameter family of functions of one variable t. Then

$$\mathcal{B}F = \{x \in R^2 : \exists t \text{ with } \partial F/\partial t = \partial^2 F/\partial t^2 = 0 \text{ at } (t,x) \text{ or } \exists$$
$$t_1 \neq t_2 \text{ with } \partial F/\partial t = 0 \text{ at each } (t_i,x), \text{ and } F(t_1,x) = F(t_2,x)\}$$

Now $\partial F/\partial t = 0$ if and only if x lies on the normal to γ at $\gamma(t)$. Thus $\mathcal{B}F$ consists of points x for which the function $F(-,x)$ has either a degenerate critical point (such x's are centers of curvature for γ, i.e., on the evolute) or two critical points at the same level (such x are the centers of bitangent circles). Thus

$$\mathcal{B}F = \text{symmetry set} \cup \text{evolute (focal set)}$$

There is a well-developed theory of full bifurcation sets (see [9]) which enables one to say what the local structure is (up to local dif-

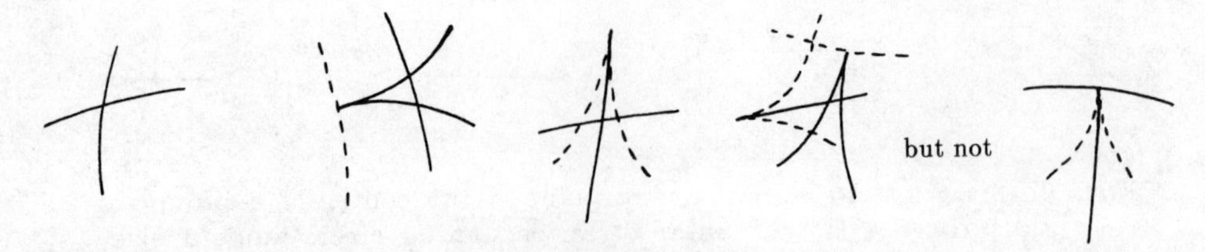

Figure 8 Self-intersections of symmetry sets.

feomorphism in R^2), given only very minimal information about the various critical points of F. In fact, F(−,x) is said to have a critical point of "type A_k" (at t) if $\partial^i F/\partial t^i(t,x) = 0$ for $1 \leq i \leq k$ and $\neq 0$ for $i = k + 1$. (See for example [7], p. 43.) It has "type $A_{k_1}A_{k_2}$" (at (t_1,t_2)) if and only if it has type A_{k_1} (at t_1) and type A_{k_2} (at t_2, where $t_2 \neq t_1$), "and $F(t_1,x) = F(t_2,x)$". "Type $A_{k_1}A_{k_2}A_{k_3}$" is similarly defined, and, for example, $A_1A_1A_1$ is abbreviated to A_1^3. Then for a generic curve γ the set $\mathcal{B}$ F is determined up to local diffeomorphism in the plane by the numbers k_i, and furthermore $\sum k_i \leq 3$. In fact, the local pictures are precisely those of Figure 7, where all cusps are ordinary cusps (locally diffeomorphic to $x^2 = y^3$). It is also possible for different local forms to intersect (Figure 8), but in general they will do so in such a way that special points (cusps, triple crossings, endpoints) do not coincide with other points of the symmetry set.

3.2 Families of Curves

What happens when special coincidences do occur? The general theory is capable of handling only generic situations, and this means that we must consider a one-parameter family of curves, say γ_u for $u \in (-1, 1) \subset R$ such that the special situation occurs for $u = 0$. The whole family then exhibits a transition. In fact, the families in Section 2.3 above, with all but one parameter held fixed, make very good experimental subjects. By studying the family of quartic function graphs $y = x^4 + bx^2$, Stephen Brassett (see [10], p. 706) discovered what we called the moth transition; see Figure 9, where only the "positive" part of the symmetry set is shown. This was before a list of all possible transitions was produced [8] by theoretical arguments, and it served as an impetus to produce that list.

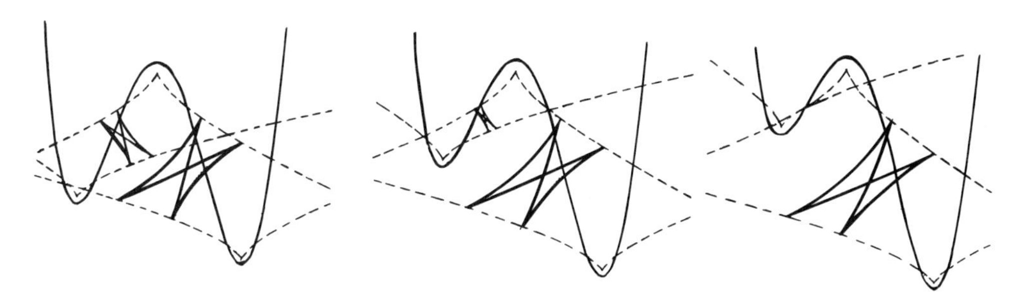

Figure 9 A moth transition on symmetry sets of an evolving curve.

The moth transition occurs when a circle is momentarily biosculating (of type A_2^2), and γ crosses the circle as in Figure 10, left. It is not too hard to believe that the circle is "rigid" in the sense that no small movement takes it to a nearby bitangent circle.

I do not have convincing pictures of many of the other transitions actually occurring in an explicit family of curves; getting the parameters just right and the transition big enough to see is an extremely delicate matter. The third part of Figure 4 actually exhibits not only the aftermath of an A_4 transition (fairly clearly) but also A_1A_3: there really is a triple crossing there!

The theoretical setup is very simple: we consider the "big family"

$$F : I \times R \times R^2 \to R$$

given by

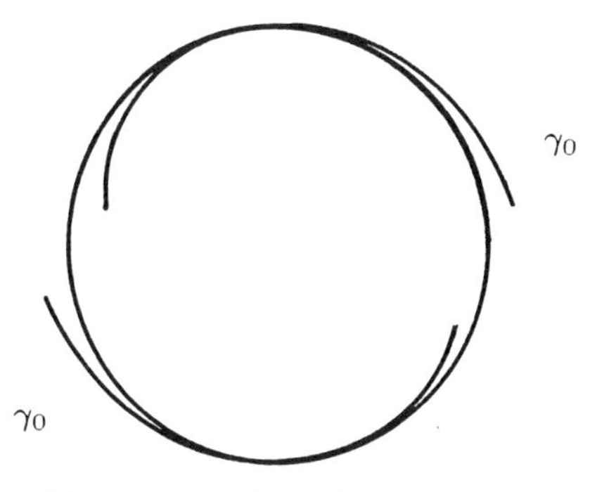

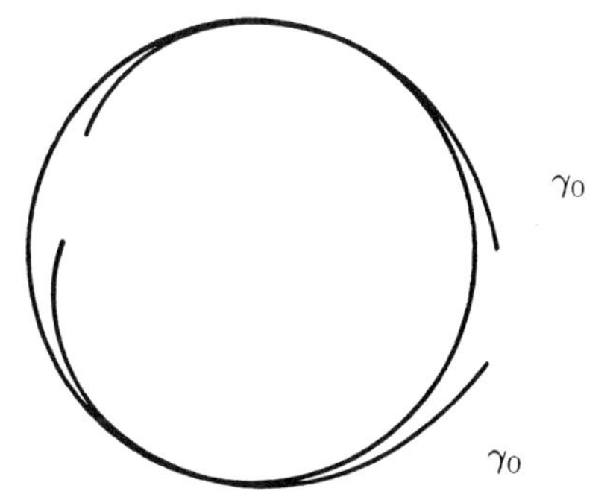

Figure 10 Biosculating circles.

$$\tilde{F}(t,u,x) = ||\gamma_u(t) - x||^2$$

We regard $\tilde{F}$ as a three-parameter family (parameters u, x_1, x_2) of functions of one variable t. Note that $\tilde{F}_u = \tilde{F}(-,u,-)$ is just F for the curve γ_u. The full bifurcation set $\mathcal{B}F \subset R \times R^2$ (coordinates (u, x_1, x_2)) then contains the individual sets $\mathcal{B}\tilde{F}_u$ for the curves γ_u as the slices u = constant. There are two problems:

1. What is $\mathcal{B}\tilde{F}$, at any rate locally?
2. How do the fibers of the projection $\mathcal{B}\tilde{F} \to R$, $(u,x) \longmapsto u$, vary as u varies?

Question 1 is relatively easy to answer; see [9]. For a <u>surface</u> M in R^3, the symmetry set consists of all centers of bitangent <u>spheres</u>, and the focal set consists of the centers of all spheres having higher tangency at some point of M. (Their centers are then principal centers of curvature for M.) Again we have a family of squared-distance functions on M (now a three-parameter family of functions of two variables, the latter being local coordinates on M) and, for this family,

Full bifurcation set = symmetry set ∪ focal set

Notice that $\mathcal{B}\tilde{F}$, though it has three parameters u, x_1, x_2, has only one variable t; however, certain singularities of functions of two variables—the corank one, or A_k singularities—are reducible to one variable without affecting their bifurcation sets. Thus, for example, the set $\mathcal{B}\tilde{F}$ which arises from a generic family γ_u for which γ_0 has type A_2^2 (a bi-osculating circle) is locally diffeomorphic to the (symmetry set ∪ focal set) for a surface having a sphere with the <u>same</u> A_2^2 contact, that is, a sphere whose center is a principal center of curvature at two points of M). A standard model for the symmetry set, up to local diffeomorphism, is the set in R^3 defined near (0,0,0) by

$$S = \{(s_1^2, s_2^2, s_1^3 + s_2^3) : s_1, s_2 \in R\}$$

See Figure 11 and [9], p. 172. The standard model for $\mathcal{B}\tilde{F}$ is the union B of S and the planes $y_1 = 0$, $y_2 = 0$ in R^3 (coordinates (y_1, y_2, y_3)).

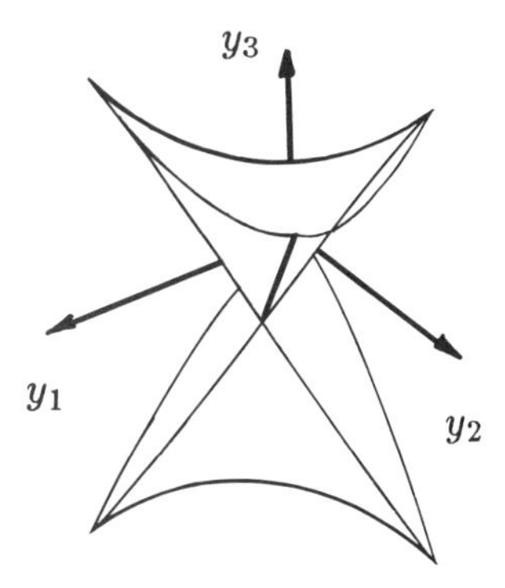

Figure 11 Symmetry set of a surface corresponding to biosculating (A_2^2) contact of circles.

When $\mathcal{B}\tilde{F}$ is transformed to this by a local diffeomorphism, the projection $\pi : R^3 \to R$, $(u,x) \longmapsto u$ is turned into some submersion $f : R^3 \to R$, defined say near (0,0,0) and taking this point to 0. We then seek a further local diffeomorphism $\phi : R^3 \to R^3$, taking the set B to itself, such that $g = f \circ \phi$ has some simple "normal form," for which the fibers $g^{-1}(u) \cap B$ are easy to describe. This family of fibers, for u increasing through 0, gives an accurate picture of the family of fibers $\pi^{-1}(u) \cap \mathcal{B}\tilde{F}$, that is, of the family of symmetry sets and evolutes of the curves γ_u.

The question is, then, what are the normal forms g? Constructing diffeomorphisms ϕ is done by integrating vector fields, and since we want $\phi(B) = B$ we must make these vector fields tangent to the singular surface B, that is, tangent to its smooth part. (It is then automatically tangent to the other lower-dimensional smooth strata, by Whitney A-regularity. For the particular surface B above, coming from type A_2^2, the singular set of B consists of two cuspidal curves ($s_1 = 0$ and $s_2 = 0$), one line ($y_1 = y_2 = 0$), and one half-line ($s_1 = s_2 \geq 0$), all meeting in the most singular stratum, namely the origin.) Methods for finding all such vector fields have been developed by several people [6,14,15], and it is often possible to find normal forms for the maps g.

There is a snag: the normal forms contain "moduli;" that is, there is a continuum of distinct possibilities. To obtain a finite list, and therefore to obtain a useful list of pictures showing the transitions, we must weaken the equivalence relation from "diffeomorphism" to "stratified homeomorphism." The practical consequence of this is that, in the pictures of transitions, smooth points, cusp points, and crossings of curves do correspond correctly to these features of the symmetry sets and evolutes—and that is enough to make the pictures visually accurate. (In fact, there is reason to believe (see [8], section 4.9 for details) that all except the transitional ($u = 0$) picture are accurate up to local diffeomorphism; there may, on the other hand, be cross-ratios in the $u = 0$ picture which will change under a stratified homeomorphism.)

To return to the example A_2^2: there are, for a generic family γ_u, two normal forms for g, namely

$$g(y_1,y_2,y_3) = y_1 + y_2 + y_3 \text{ and } g(y_1,y_2,y_3) = y_1 - 2y_2 + y_3$$

The sections g = constant of B give respectively the moth transition and another transition which occurs when the biosculating (A_2^2) circle has the configuration of Figure 10, right.

Figure 12 shows the complete list of generic transitions on one-parameter families of symmetry sets and evolutes (the latter drawn dashed) in the plane. See [8]. Note that for evolutes the <u>only</u> generic transition is the "swallowtail," which can be seen in the A_4 picture of Figure 12. Thus a considerable richness of detail is added by considering also the symmetry set.

Note also that, although the number of cusps on the symmetry set always changes by an even number (either 0, 2, or 4) in a transition, the number of triple crossings can change by 1 (in the A_1A_3 transition). See Section 4 for a relation connecting triple crossings, cusps, and endpoints on the symmetry set.

3.3 The Piecewise-Linear Case Again

It is a striking fact that nearly all the transitions of Figure 12 have close analogs in the simplest possible situation of families of piecewise-linear (PL) curves (compare Section 2.4 above) and that they <u>all</u> have analogs for piecewise-circular (PC) curves. (A PC curve is formed from a sequence of circular arcs and straight lines in such a way that

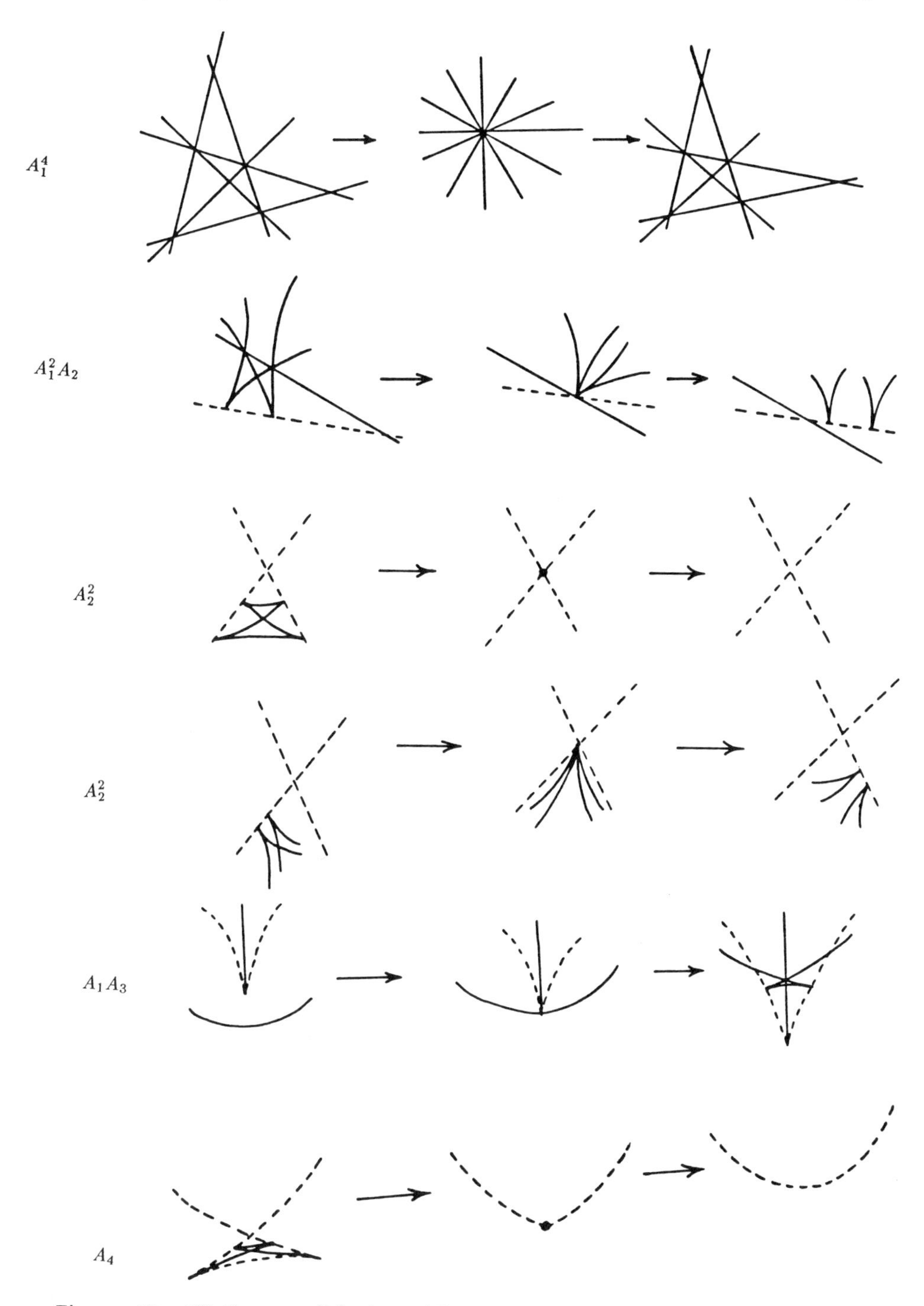

Figure 12 All the possible transitions on symmetry sets in a generic one parameter family of plane curves.

the directed tangent line varies continuously along the curve.) Figure 6, left, shows the beginning of a moth transition in a PL family, but there does not seem to be a sensible analog of A_4 circles for PL curves, so that the A_4 transition cannot be convincingly imitated. Full details of the PL and PC situations will appear elsewhere, in joint work with T. F. Banchoff. From a computer graphics standpoint, the construction of symmetry sets for PC curves represents a considerably greater challenge than is the case for polygons. The first task, of course, is to consider the contribution from a pair of circular arcs. It is a pleasant exercise to show that this consists of one or more parts, each of which is an arc of an ellipse, hyperbola, or parabola.

4. A GLOBAL RESULT

4.1 Plane Curves and Spherical Curves

We can apply a result of Ozawa [13] concerning triple tangencies of space curves to the study of symmetry sets in the plane. In fact, any plane curve γ can be turned by inverse stereographic projection to a space curve $\bar{\gamma}$ which happens to lie on a sphere. Multiple-contact circles for γ lift to <u>circles</u> on the sphere which are cut out by multiple-contact planes for $\bar{\gamma}$. In fact, we have the following correspondence:

$\bar{\gamma}$	$\leftrightarrow$	γ	
Tritangent plane	$\leftrightarrow$	Tritangent circle	(A_1^3)
Plane tangent at one point and osculating at another	$\leftrightarrow$	Circle tangent at one point and osculating at another	(A_1A_2)
Osculating plane at a simple torsion zero	$\leftrightarrow$	Circle osculating at a vertex	(A_3)

Now by suitably indexing the planes of these three types (on the left in the table of correspondences), Ozawa proved two integer relationships between the sums of indices for the three types.

4.2 Indices for Bitangent Circles

Here is the interpretation of the Ozawa indices for symmetry sets of simple closed plane curves γ. Every circle, and also γ, is oriented anticlockwise (that is, with interior on the left), and each reversal of an underlined word below changes the sign of the index.

1. Consider an A_1^3 (tritangent) circle (the center is a triple crossing on the symmetry set). If the three points of contact in order on γ give the <u>anticlockwise</u> orientation to the circle and if γ is locally exterior to the circle for an <u>odd</u> number of points of tangency, then the index is +1. Let T denote the sum of indices over all such circles.
2. Consider an A_1A_2 circle (the center is a cusp on the symmetry set). If the curve γ close to the osculation point has orientation <u>agreeing</u> with that of the circle, and if γ is locally <u>exterior</u> to the circle at the other point of tangency, then the index is +1. Let C be the sum of the indices of all such circles.
3. Consider an A_3 circle (the center is an endpoint on the symmetry set). If the orientations of γ and the circle <u>agree</u> at the point of contact and γ is locally <u>interior</u> to the circle at this point, then the index is +(the number of transverse intersections of the circle with γ). Let E denote the sum of these indices. (Unlike those in 1 and 2, each index can in principle be an arbitrary integer.)

Ozawa's theorem in [13] then gives us:

Theorem

$C = E = 2T$

As an example, suppose that γ has exactly four vertices (the smallest possible number). Then it is not hard to show that the circle of curvature at a vertex (an A_3 circle) cannot meet γ again (see [11], p. 576), so that E = 0 (note that this does <u>not</u> assert that there are no A_3 circles!). Hence C = T = 0, and in particular if there is a tritangent (A_1^3) circle then there must be another one of opposite index. But I do not know whether a four-vertex curve can have tritangent circles at all.

It is possible to trace the indices directly through all the transitions of Section 3.2 and to verify directly that C, E, and T are preserved by each transition [2]. This gives a much clearer indication why these indices are the appropriate ones to take. It is also possible to assign indices to appropriate multiply tangent (or multisupport) circles for PL and PC curves in such a way that the relationships of the theorem still hold. Surprisingly, this is even possible for A_3 circles in the case of PL curves. See [1,2].

ACKNOWLEDGMENT

The work reported here was done jointly with T. F. Banchoff, S. A. Brassett, J. W. Bruce, and C. G. Gibson. The chapter was written while I was a visiting professor at the University of Massachusetts at Amherst, and I am grateful to Five Colleges, the University of Massachusetts, the University of Liverpool and Brown University for financial support.

REFERENCES

1. T. F. Banchoff and P. J. Giblin, Bisupport circles for planar polygons, in preparation.
2. T. F. Banchoff and P. J. Giblin, Global theorems for symmetry sets of smooth curves and polygons in the plane, Proc. R. Soc. Edinburgh Sect. A, 106, (1987), 221–231.
3. H. Blum, Biological shape and visual science, I, J. Theor. Biol. 38, (1973), 205–287.
4. F. Bookstein, The line-skeleton, Comput. Graph. Image Process., 11, (1979), 123–137.
5. M. Brady, Criteria for representations of shape, Human and Machine Vision (J. Beck et al., eds.), Academic Press, New York (1983).
6. J. W. Bruce, Vector fields on discriminants and bifurcation varieties, Bull. London Math. Soc., 17, (1985), 257–262.
7. J. W. Bruce and P. J. Giblin, Curves and Singularities, Cambridge University Press, Cambridge (1984).
8. J. W. Bruce and P. J. Giblin, Growth, motion and one-parameter families of symmetry sets, Proc. R. Soc. Edinburgh Sect. A, 104, (1986), 179–204.

9. J. W. Bruce, P. J. Giblin, and C. G. Gibson, Symmetry sets, Proc. R. Soc. Edinburgh Sect. A, 101, (1985), 163–186.

10. P. J. Giblin and S. A. Brassett, Local symmetry of plane curves, Am. Math. Monthly, 92, (1985), 689–707.

11. S. B. Jackson, Vertices of plane curves, Bull. Am. Math. Soc., 50, (1944), 564–578.

12. D. T. Lee, Medial axis transformation of a planar shape, IEEE Trans. Pattern Anal. Mach. Intell., 4, (1982), 363–369.

13. T. Ozawa, The number of triple tangencies of smooth space curves, Topology, 24, (1985), 1–13.

14. H. Terao, The bifurcation set and logarithmic vector fields, Math. Ann., 263, (1983), 313–321.

15. V. M. Zakalyukin, Bifurcation of fronts and caustics, depending on one parameter, Itogi Nauki, 22, (1983), 53–93.

8

On Crossing the Boundary of the Mandelbrot Set

IVAN HANDLER

Independent Software Developer
Chicago, Illinois

LOUIS H. KAUFFMAN and DAN SANDIN

University of Illinois,
Chicago, Illinois

This is a summary of our computer-exploratory work investigating qualitative changes in the Julia set of the functions of the form

$$f(z) = z^n + c$$

as the complex parameter, c, moves across the boundary of the Mandelbrot set.

Background for this exploration is provided in [1]. A review of this material is included:

Definition 1

In the case that f is a polynomial, the Julia set of f is the frontier of $\{z : f^n(z) \to \infty;\ \text{as } n \to \infty\}$. The Fatou set is the complement of the Julia set. A more sophisticated definition is required for nonpolynomial mappings. For this we refer the reader to [1].

Note. It is clear from this definition that the Julia set of $f(z) = z^2$ is the unit circle in the complex plane.

Theorem 2

If the number of components of the Fatou set is finite, then there are at most two such components.

Definition 3

The Mandelbrot set of the function $f_{n,c} = z^n + c$ is the set of all c such that the Julia set of $f_{n,c}$ is connected.

Remarkably, there is a simple algorithm to determine (for fixed n) whether $f_c = f_{n,c} = z^n + c$ has a connected Julia set. The criterion is given by:

Theorem 4 (Fatou and Julia)

The Julia set of $f_c = z^n + c$ is totally disconnected if and only if

$$\lim_{k \to \infty} f_c^k(0) = \infty$$

Otherwise the Julia set is connected.

Using the iteration

$$f_c(0),\ f_c f_c(0),\ f_c f_c f_c(0),\ \ldots$$

we quickly determine the connectivity of the Julia sets that we graph. Thus, this theorem gives information that would be at best conjectural from the computer graphics alone. We really do know (Platonically!) the connectivity of the sets that we graph. (Of course, for certain cases very near the boundary of the Mandelbrot set, the iteration may not diverge in a human lifetime, but we _have_ checked all cases up to 30,000 iterations.)

As a consequence of this theorem, it is possible to produce pictures of the set M_n of those c such that $f_{n,c}$ has a connected Julia set. The set M_n is called the Mandelbrot set. Benoit Mandelbrot was the first person to distinguish and investigate this set using computer graphics [2]. This renewal of research via computers marks a new stage in the investigation of these sets.

In the case that the Fatou set of $f_{n,c}$ has one component, either the Julia set is totally disconnected and the constant c is in the exterior of the Mandelbrot set or the Julia set is a dendrite—locally connected—and c is an element of the Mandelbrot set. The Fatou set of $f_{n,c}$ cannot have zero components. The transition from an F of two components to an F of one component can be quite different from the transition from an F of an infinite number of components to one of one component (see Figures 8 to 24).

Figures 1 to 7 demonstrate the transition from a totally disconnected Julia set just up to the boundary of the Mandelbrot set. The constant c is changing by 0.001i at each step (after the first). Some changes happen rather quickly, especially in the last three steps from c = 0.325 + 0.05i to 0.325 + 0.052i. Notice that the Fatou set of the final stage has two components.

The figures were generated using a algorithm based on the fact that the Julia set is contained in the closure of the inverse images of the iterates of $f_{n,c}$ (corollary 4.7 in [1]). This algorithm is fast at producing images at the cost of lack of uniformity—some areas of the Julia set are more filled in than others.

Figures 8 to 12 demonstrates a transition from a Julia set whose associated Fatou set has an infinite number of components.

Figures 13 to 24 demonstrate transitions from Julia sets with associated Fatou sets with two components to totally disconnected Julia sets back to Julia sets with associated Fatou sets with an infinite number of components (only a portion of the whole sequence is shown). In Figures 13 to 22, $f_{n,c}$ is quadratic, and in Figures 23 and 24, $f_{n,c}$ is cubic. What is interesting here is that we see spiraling into what looks like limit cycles. These cycles become smaller and smaller and finally become the vertices of the Julia polygons that enclose the components of the Fatou set. While the spiraling was also noticed in the first set of pictures, we have not (as yet) found any Julia sets nearby where the same type of transition to polygons occurs.

This raises some questions: How can we characterize this spiraling and when does it occur? Under what circumstances will this spiraling behavior end with the polygons and when will it end with Fatou sets with only two components?

We also illustrate how the qualitative change of form may in certain cases be modeled by a simple combinatorial change in the generator for a directly recursively generated fractal. See Figure 25 and [2]. Some questions arise from these pictures: Can a connected Julia set be de-

Figure 1

Figure 2

Figure 3

Figure 4

Figure 5

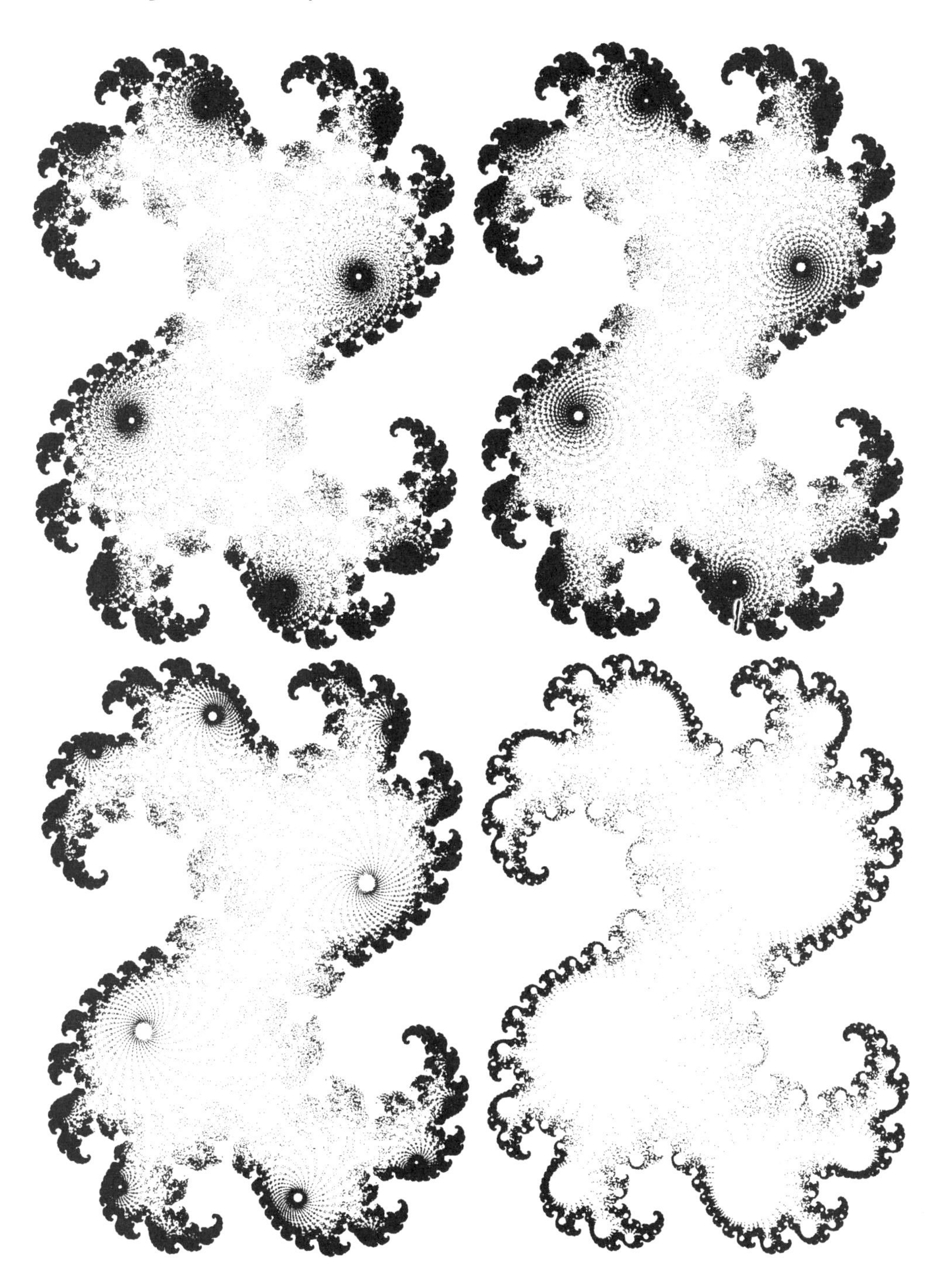

Figure 6

Figure 7

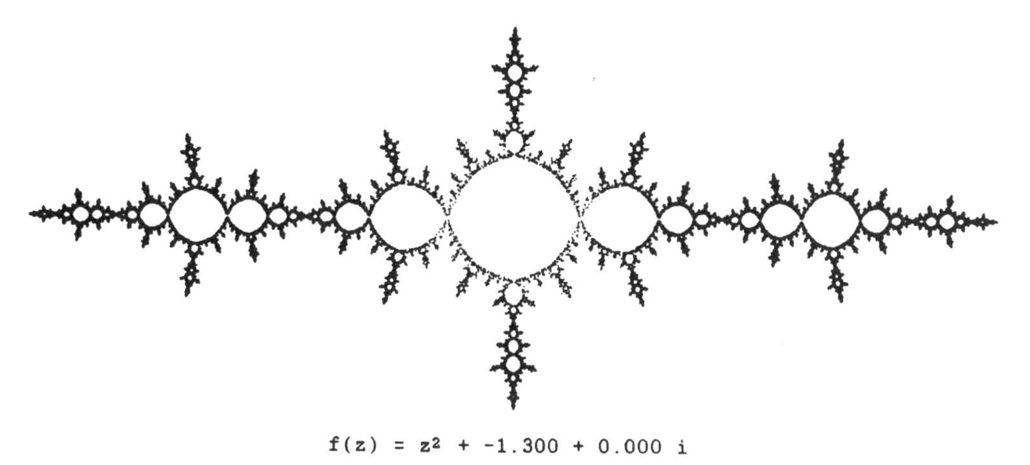

$f(z) = z^2 + -1.300 + 0.000\ i$

Figure 8

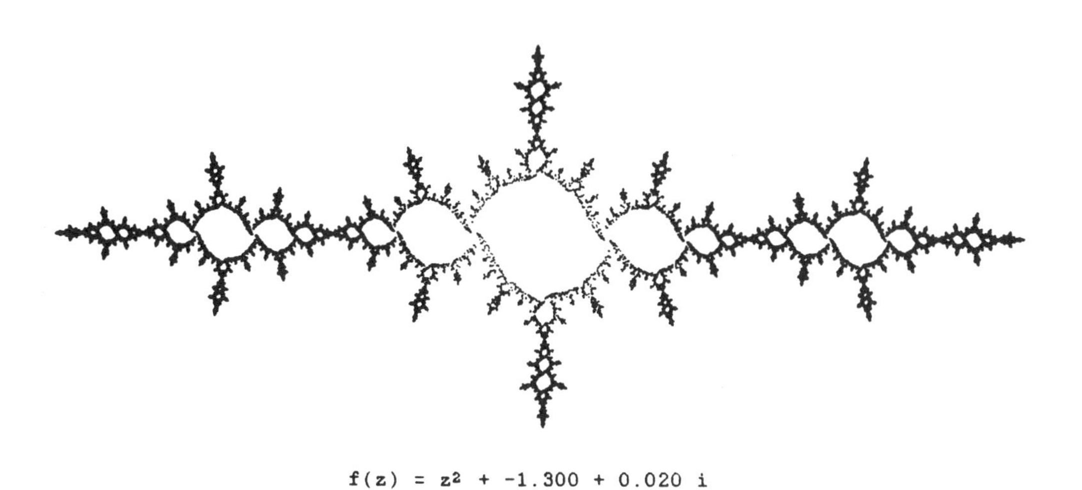

$f(z) = z^2 + -1.300 + 0.020\ i$

Figure 9

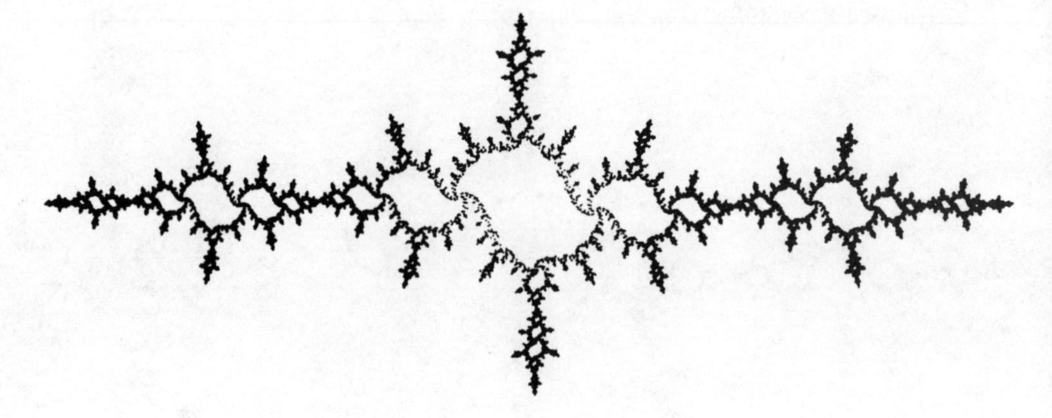

$f(z) = z^2 + -1.300 + 0.040\ i$

Figure 10

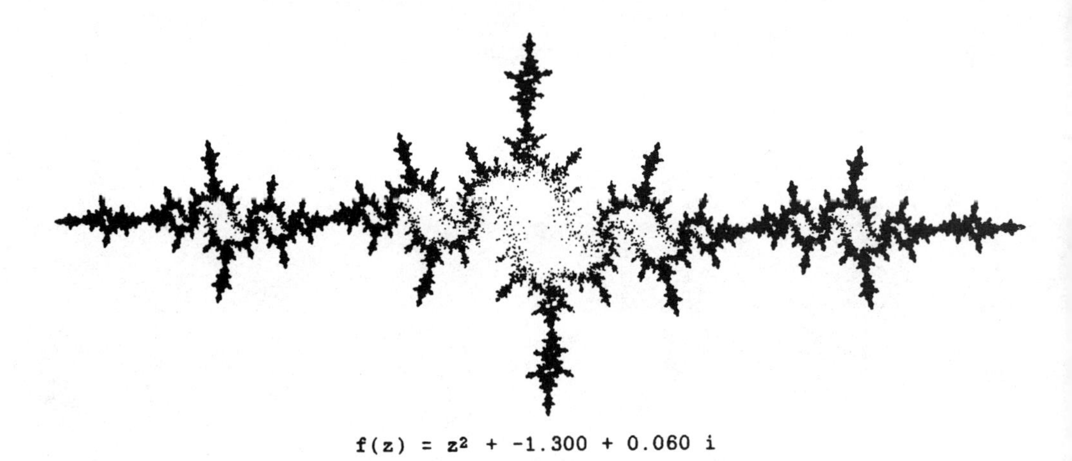

$f(z) = z^2 + -1.300 + 0.060\ i$

Figure 11

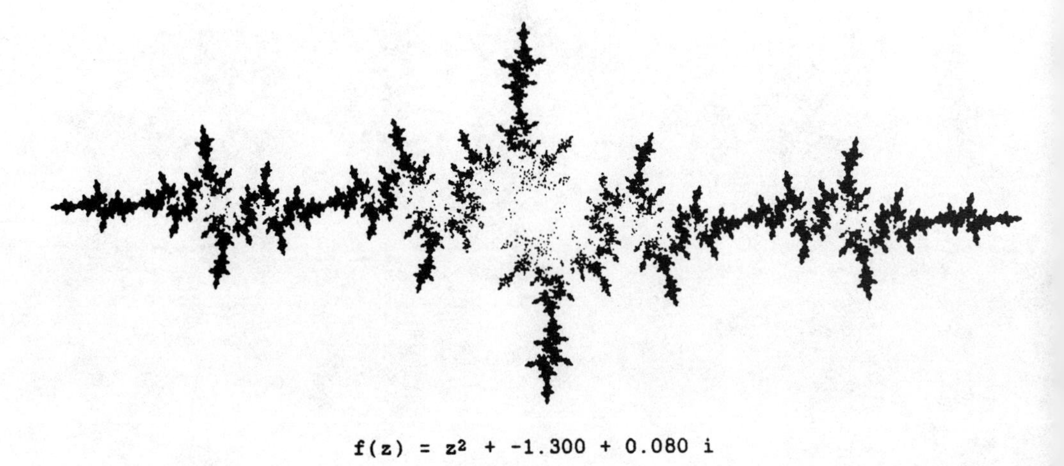

$f(z) = z^2 + -1.300 + 0.080\ i$

Figure 12

$f(z) = z^2 + 0.300 + 0.570\,i$

Figure 13

$f(z) = z^2 + 0.300 + 0.571\,i$

Figure 14

$f(z) = z^2 + 0.300 + 0.572\,i$

Figure 15

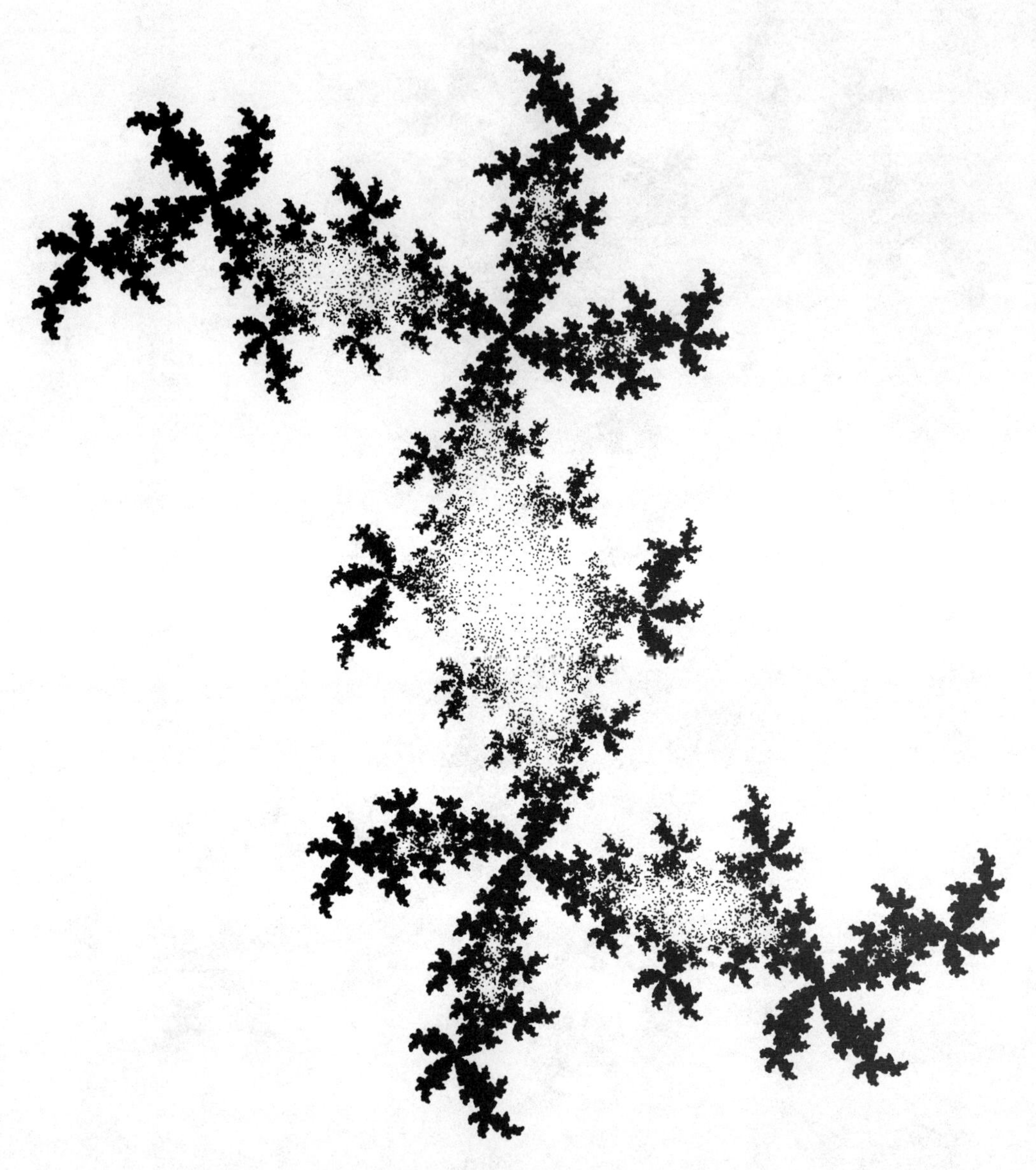

$f(z) = z^2 + 0.300 + 0.573\,i$

Figure 16

$f(z) = z^2 + 0.300 + 0.574\,i$

Figure 17

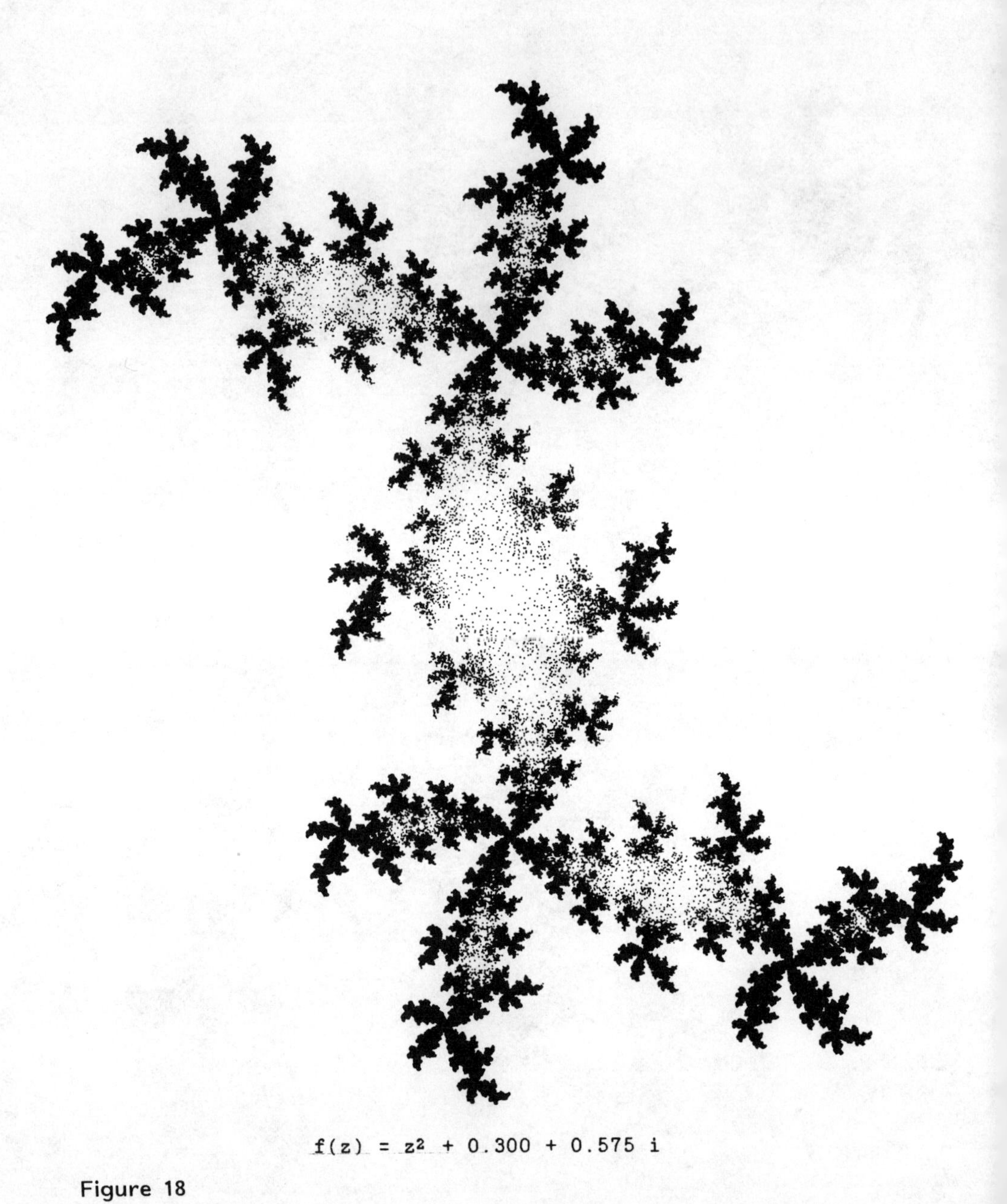

$f(z) = z^2 + 0.300 + 0.575\, i$

Figure 18

$f(z) = z^2 + 0.300 + 0.576\,i$

Figure 19

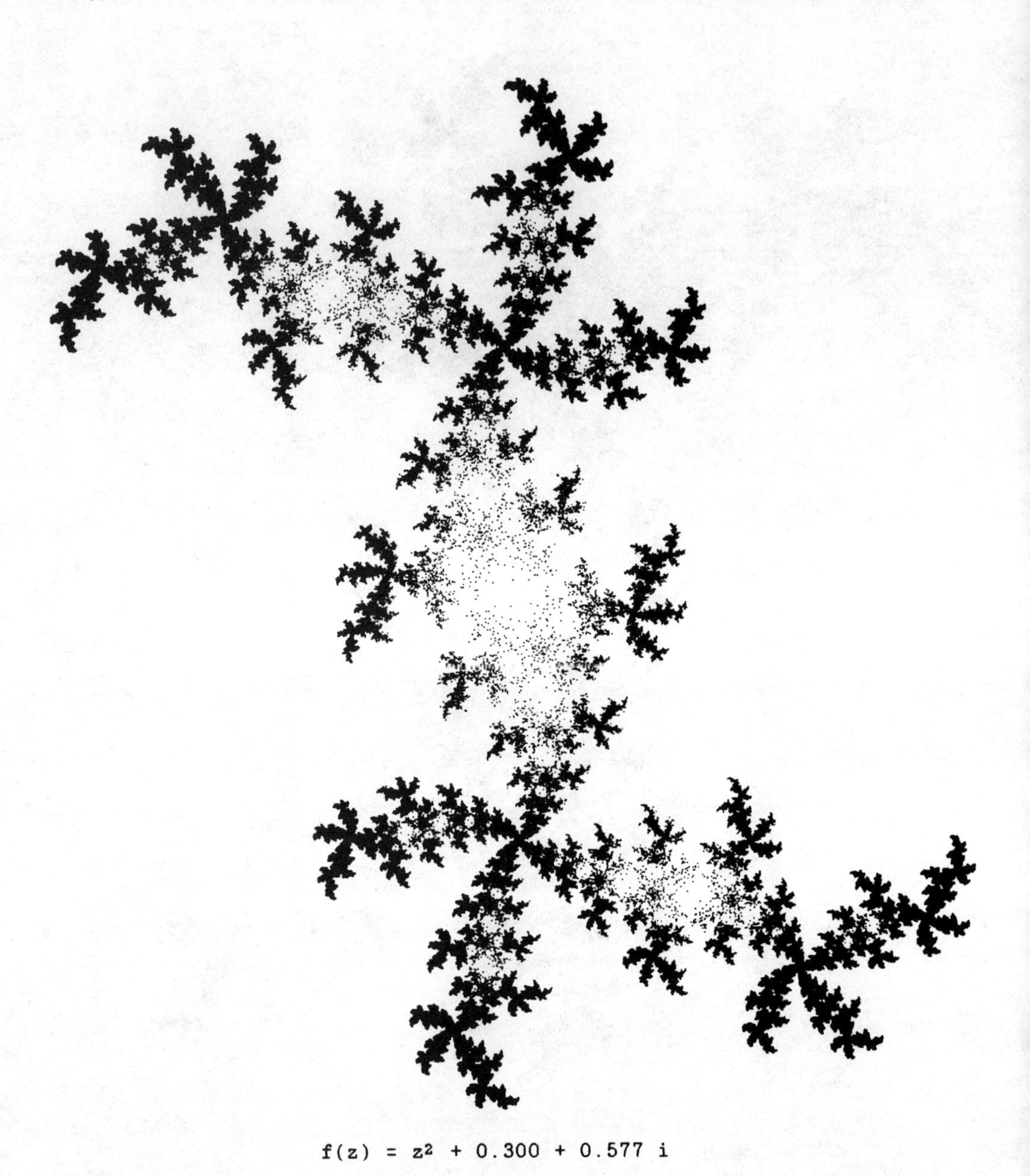

$f(z) = z^2 + 0.300 + 0.577\, i$

Figure 20

$f(z) = z^2 + 0.300 + 0.578\,i$

Figure 21

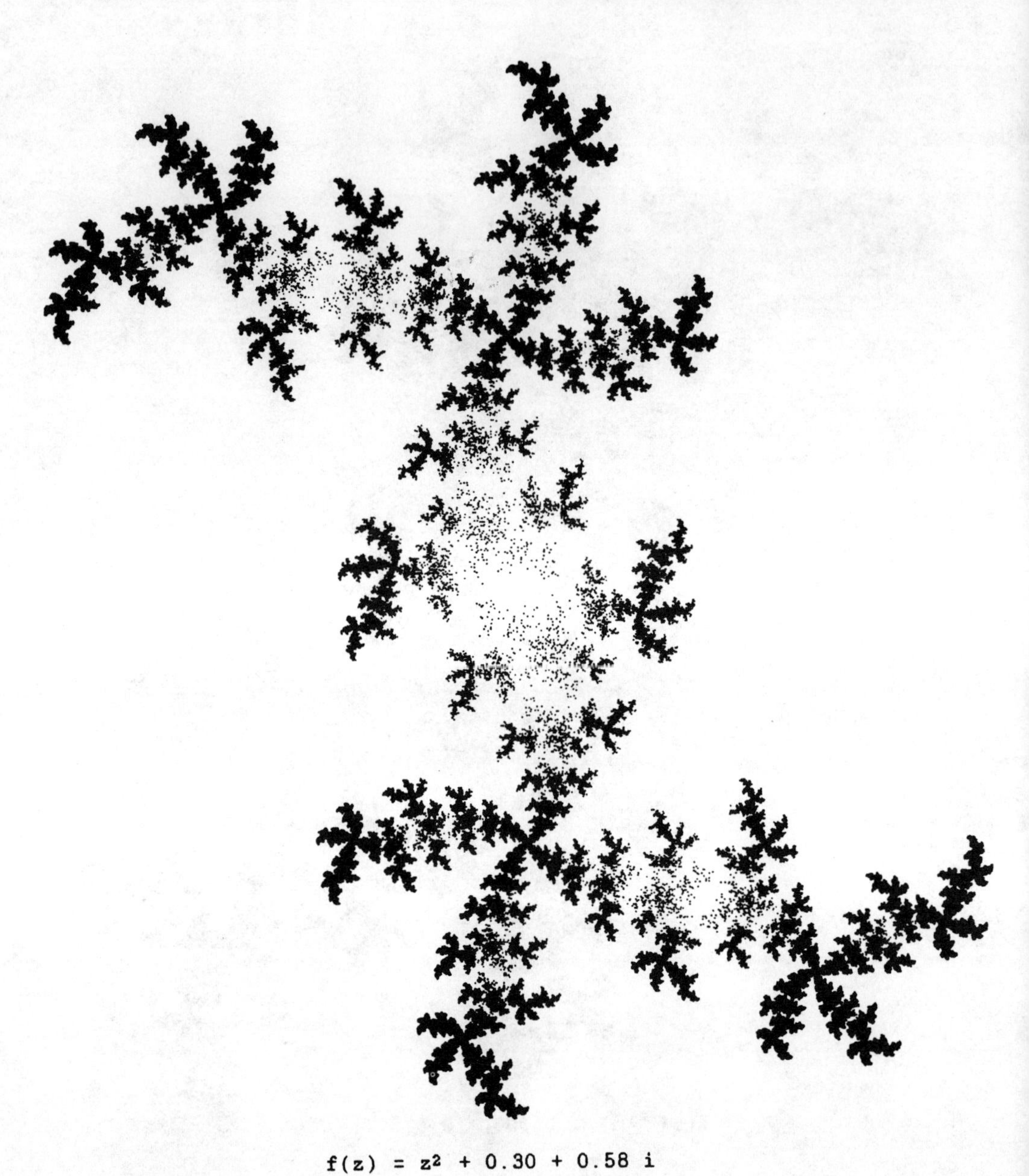

$f(z) = z^2 + 0.30 + 0.58\,i$

Figure 22

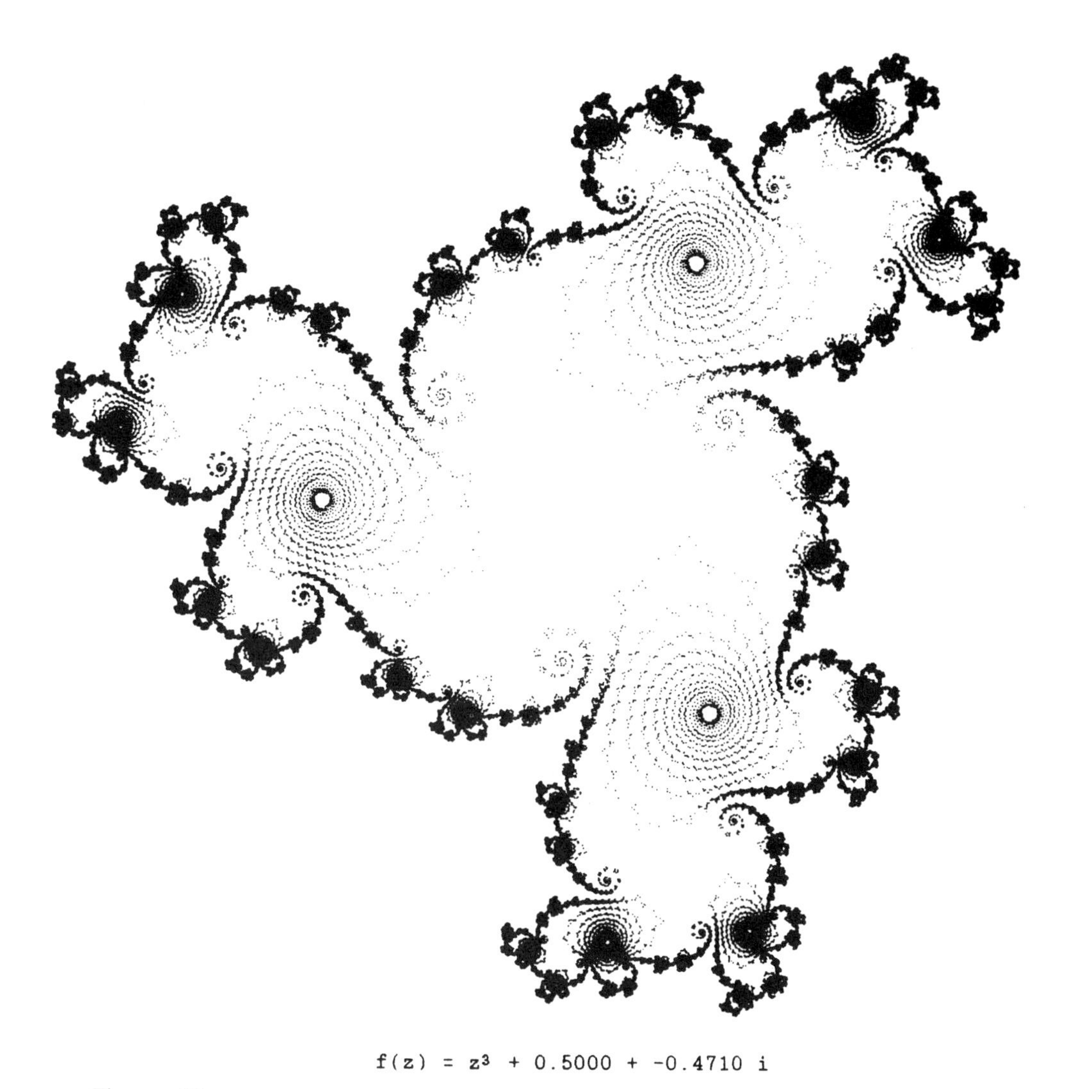

$f(z) = z^3 + 0.5000 + -0.4710\,i$

Figure 23

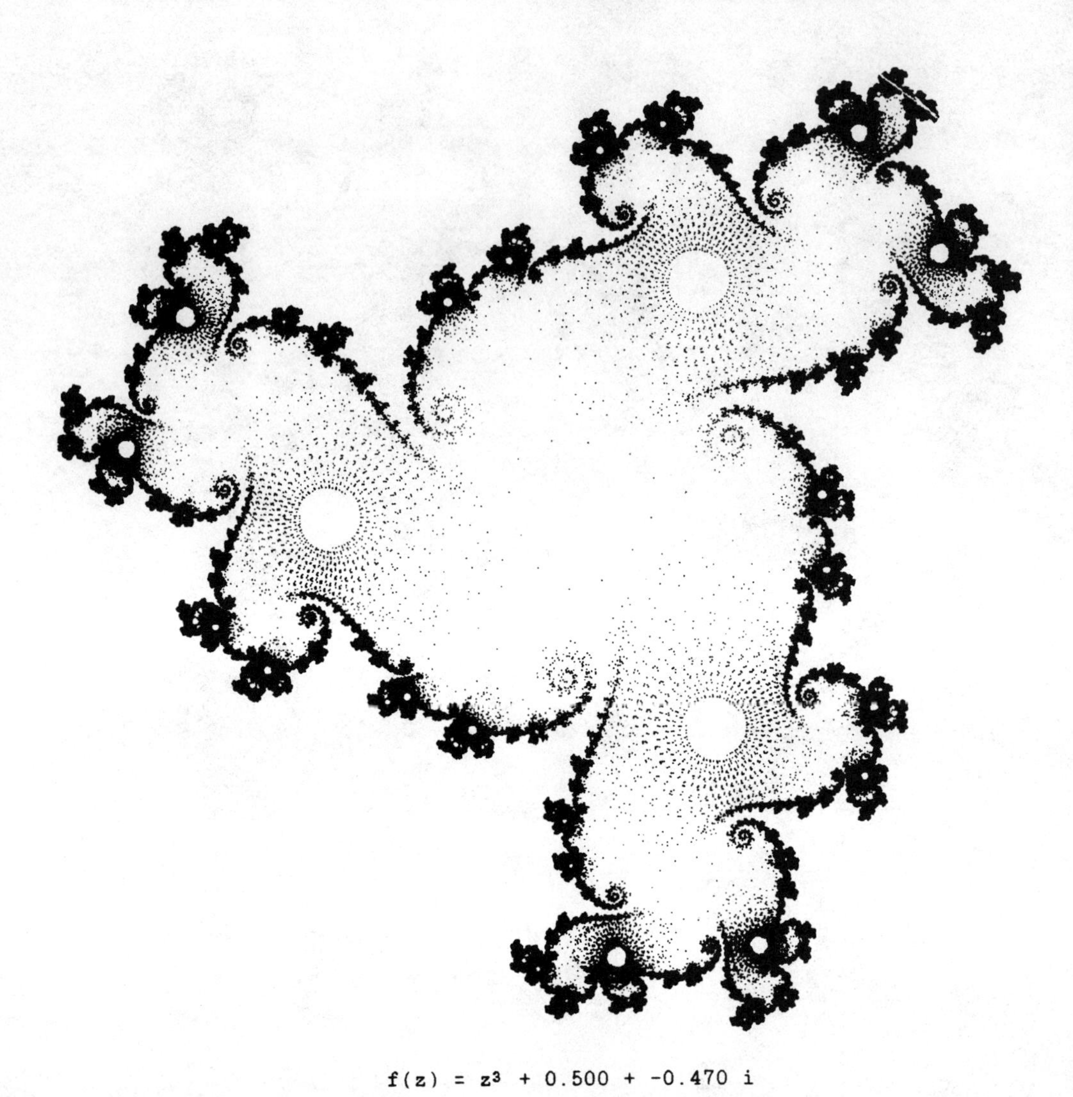

$f(z) = z^3 + 0.500 + -0.470\ i$

Figure 24

Figure 25

scribed by such a process? How can one derive estimates for the sizes of features of the recursive form? This last question would be useful for computing the fractal dimension of the Julia set. Is there an analog to the Cantor middle third process that can describe a totally disconnected Julia set?

Note on the algorithm. Milnor suggested to one of the authors (Kauffman) a modification of the Julia set algorithm. The original algo-

Figure 26

rithm recursively computes inverse images of points until the point to be plotted has already been plotted (a collision). Milnor suggested allowing the algorithm to continue until some (previously specified) number of collisions occurs. This modified algorithm can enhance these images. For example, Figure 26 was generated using the enhanced algorithm. It was generated using a collision number of 6 and took over 24 hours to compute. This modification works because a collision means only that two points are within a pixel's diameter of each other. If you keep on going down the tree of inverse images of two points that collide, they may diverge farther down, revealing more detail.

The images were produced on a Compaq microcomputer with an 8087 coprocessor and an Epson FX-85 printer. The program was written in Microsoft Pascal version 3.31 and needs at least 400K. It takes between 10 and 40 minutes to produce an image (at collision number 1). Each image is printed on an 8-inch-square area. The program is available from the first author on a $5\frac{1}{4}$-inch disk. A program in BASIC on cassette, requiring 10K, for the Model 100 is available from the second author. Other graphics materials related to Julia sets are available from Dan Sandin.

REFERENCES

1. P. Blanchard, Complex analytic dynamics on the Riemann sphere, Bull. Am. Math. Soc., (July 1984), 85.
2. B. Mandelbrot, The Fractal Geometry of Nature, Freeman, San Francisco (1982).

9

A Stable Decomposition of BSD_{16}

JOHN C. HARRIS*

Purdue University Calumet
Hammond, Indiana

1. INTRODUCTION

Let BG be the classifying space of a finite p-group, G. Consider the problem of finding a stable decomposition

$$BG \simeq X_1 \vee X_2 \vee \cdots \vee X_N$$

into indecomposable wedge summands, localized at p. It is known that such wedge summands correspond to primitive idempotents in the ring $\{BG,BG\} \otimes Z/p$, the mod-p stable self-maps of BG. A Krull-Schmidt theorem implies that any two decompositions are homotopy equivalent, so to solve the topological problem, it suffices to find just one collection of primitive orthogonal idempotents in $\{BG,BG\} \otimes Z/p$ which sum to the identity. Unfortunately, it is generally difficult to find idempotents and, given one, it is difficult to determine whether or not it is primitive.

From the explicit algebraic description of $\{BG,BG\}$ given by May [4], it is easy to see that $\{BG,BG\} \otimes Z/p$ is finite. Hence one might list all of the idempotents and pick out a set of primitive orthogonal ones that sum to the identity. This method was used in [2] to decompose BZ/p, $B(Z/2 \times Z/2)$, and $B(Z/3 \times Z/3)$. For G any abelian p-group, the complete decomposition of BG has been described in terms of

*Current affiliation: University of Washington, Seattle, Washington

representation theory [3]. When $p = 2$, the decompositions for the dihedral and generalized quaternion groups were found in [6].

In this chapter the above method is used to decompose the classifying space of $G = SD_{16}$, the semidihedral group of order 16. It is necessary to generate the list of idempotents on a computer—$\{BSD_{16}, BSD_{16}\} \otimes Z/2$ has 2^{51} elements.

The chapter is organized as follows. In Section 2 we recall some facts about stable wedge summands and May's description of $\{BG, BG\} \otimes Z/p$. Section 3 collects some general results which we will need. The list of idempotents in $\{BSD_{16}, BSD_{16}\} \otimes Z/2$ is found in Section 4. From the list it can be shown that the identity decomposes into four primitive idempotents. We identify all but one of the summands in Section 5 using homological calculations. Our main result is the following.

Proposition 5.5

$BSD_{16} \simeq BZ/2 \vee \Sigma^{-1}(BS^3/BN) \vee L(2) \vee X$ is a complete decomposition.

Throughout, p will denote a fixed prime, $H^*(\_)$ will denote $\tilde{H}^*(\_;Z/p)$, and all spectra will be localized at p. For X and Y spaces or spectra, $\{X, Y\}$ will denote the stable maps from X to Y, and X_+ will mean X with a disjoint basepoint.

After completing this research, the author learned that a more general result has been found by John Martino in his thesis at Northwestern University. He finds complete decompositions for BSD_{2^n} for $n \geq 4$. It follows from his results that the summand X in Proposition 5.5 is homotopy equivalent to $BPSL_3(3)$, localized at 2.

2. REDUCTION TO ALGEBRA

It is well known that wedge summands of a spectrum X correspond to idempotents in the ring $\{X, X\}$—the summand associated to an idempotent $e \in \{X, X\}$ is denoted eX, and $X \simeq eX \vee (1 - e)X$. (The eX can be defined either by Brown representability, $\{Y, eX\} = e\{Y, X\}$, or more directly as the telescope $\mathrm{Tel}(X \xrightarrow{e} X \xrightarrow{e} X \cdots)$.) It follows that $H^*X \cong H^*(eX) \oplus H^*((1 - e)X)$ is an isomorphism of modules over the Steenrod algebra and that $H^*(eX) \cong e^*(H^*X)$.

When X = BG, the classifying space of a finite p-group, it can be shown that the wedge summands correspond to idempotents in $\{BG,BG\} \otimes Z/p$ (i.e., one can work modulo p) [3]. So to find a complete wedge decomposition of BG it suffices to find primitive orthogonal idempotents $\{e_1,e_2,\ldots,e_N\}$ in $\{BG,BG\} \otimes Z/p$ which sum to the identity.

The ring $\{BG,BG\} \otimes Z/p$ can be described algebraically [4]. We first define the generalized Burnside ring, A(G,G). Consider the set of all homomorphisms from subgroups of G to G. There is an equivalence relation (~) on these homomorphisms given by $\rho \sim \sigma$ iff $\rho = c_{g_1}\sigma c_{g_2}$, where g_1 and g_2 are in G and c_g denotes conjugation. Let $\bar{A}(G,G)$ denote the set of equivalence classes under ~ and let A(G,G) be the free abelian group on $\bar{A}(G,G)$. Define the product of two basis elements by the formula

$$\sigma * \rho = \sum_{\{x\}} \psi_x \qquad [2.1]$$

where $\{x\}$ runs over a set of double coset representatives of $\mathrm{Im}\,\rho\backslash G/\mathrm{Dom}\,\sigma$ and ψ_x is the composition $\sigma \circ c_x \circ \rho$ from $\rho^{-1}(x(\mathrm{Dom}\,\sigma)x^{-1})$ to G. Extend this linearly to all of A(G,G).

There is a natural map ε from A(G,G) to A(G), the usual Burnside ring, given on basis elements by $\varepsilon(\rho) = G/\mathrm{Dom}\,\rho$. Let K(G,G) be the kernel of ε.

Define a map $\alpha : \bar{A}(G,G) \to \{BG_+,BG_+\}$ by letting

$$\alpha(\rho) = B\rho_+ \circ tr : BG_+ \to BH_+ \to BG_+$$

where $\rho : H \to G$ in an element in $\bar{A}(G,G)$ and tr is the stable transfer. The induced map $\alpha : A(G,G) \to \{BG_+,BG_+\}$ is a ring homomorphism and restricts to give a map $\alpha : K(G,G) \to \{BG,BG\}$. These maps become isomorphisms after completing at p [4, Corollary 15]. This implies the following.

Proposition 2.2

The maps $\alpha : A(G,G) \otimes Z/p \to \{BG_+,BG_+\} \otimes Z/p$ and $\alpha : K(G,G) \otimes Z/p \to \{BG,BG\} \otimes Z/p$ are isomorphisms.

We give another description of $K(G,G)$. Let $A_e(G,G)$ denote the subgroup of $A(G,G)$ generated by the trivial homomorphisms $\rho : H \to \{e\} \subseteq G$. It turns out that this subgroup is an ideal and the composition

$$K(G,G) \subseteq A(G,G) \to A(G,G)/A_e(G,G)$$

is a ring isomorphism. So we can think of $K(G,G)$ either as a subring or as a quotient ring of $A(G,G)$. When we speak of the basis elements of $K(G,G)$, we mean those basis elements of $A(G,G)$ that are not in $A_e(G,G)$.

3. PREVIOUS RESULTS

Here we recall some general results which we will use later. Let $\overline{\mathrm{Aut}}(G) = \mathrm{Aut}(G)/\sim$ and $\overline{\mathrm{End}}(G) = \mathrm{End}(G)/\sim$. There are inclusions of rings

$$Z/p[\overline{\mathrm{Aut}}(G)] \subseteq Z/p[\overline{\mathrm{End}}(G)] \subseteq A(G,G) \otimes Z/p$$

where $Z/p[\_]$ denotes the group (or semigroup) ring.

Proposition 3.1 [3, Cor 2.8; 7, Cor 5.9]

If G is abelian and if e is a primitive idempotent in $Z/p[\mathrm{End}(G)]$, then e is primitive when considered as an idempotent in $A(G,G) \otimes Z/p$.

It follows that a primitive orthogonal idempotent decomposition of the identity in $Z/p[\mathrm{End}(G)]$ solves the splitting problem when G is abelian.

Proposition 3.2 [7, Lemma 5.1]

If X is an indecomposable summand of BG, then there is a unique (up to isomorphism) minimal subgroup H of G with X homotopy equivalent to a summand of BH.

We will say that such a summand X is associated to H. A summand associated to G is called dominant.

Proposition 3.3 [7, Theorem 5.2]

Let $\overline{BG}$ be the wedge sum of the dominant summands of BG. A primitive orthogonal idempotent decomposition $1 = e_1 + e_2 + \cdots + e_n$ in $Z/p[\overline{Aut}(G)]$ induces a complete wedge decomposition of $\overline{BG}$.

The above two propositions suggest an inductive approach to the splitting problem. Given a finite p-group G, first determine all the indecomposable summands of the classifying spaces of its proper subgroups. Then determine which of these split off BG and with what multiplicities. Finally, decompose the remaining summand, $\overline{BG}$, using the representation theory of $\overline{Aut}(G)$.

4. IDEMPOTENTS IN $K(G,G) \otimes Z/2$ FOR $G = SD_{16}$

Let $G = SD_{16}$ for the remainder of this chapter. In this section we describe how a primitive orthogonal idempotent decomposition of the identity in $K(G,G) \otimes Z/2$ was found.

From the presentation $G = \langle a,b \mid a^8 = b^2 = 1, bab = a^3 \rangle$ it is easy to show that G has a total of 15 subgroups, and with a little perseverance one can write down the 279 homomorphisms $\rho : H \to G$ with $H \subseteq G$. We are interested in equivalence classes of such objects—there are 10 conjugacy classes of subgroups and 61 equivalence classes of homomorphisms (under ~). Hence $A(G,G) \otimes Z/2$ has 2^{61} elements and $K(G,G) \otimes Z/2$ has 2^{51} elements.

Let $A_1, \ldots, A_{51}$ be the basis elements for $K(G,G) \otimes Z/2$. Using the double coset formula (2.1), the products $A_i * A_j$ can be found. An element $\sum a_i A_i$, $a_i \in Z/2$, will be idempotent iff

$$\sum a_i A_i = \sum \sum a_j a_k A_j * A_k \qquad [4.1]$$

By equating the coefficients of the A_i, we obtain 51 equations in the unknowns $a_1, \ldots, a_{51}$. To solve these equations we could just try each of the possible choices for the a_i's.

We reduce the number of equations to solve as follows. First, assume A_{51} represents the identity automorphism of G. Then an element e in $K(G,G) \otimes Z/2$ is an idempotent iff $A_{51} + e$ is. Hence we can assume $a_{51} = 0$ and find half of the idempotents.

It follows from the $A_j * A_k$ multiplication table (under the assumption that $a_{51} = 0$) that some basis elements do not occur on the right side of (4.1), so their coefficients are zero. Also, some of the equations have the same right sides, so their coefficients are equal. At this point we are reduced to 37 equations, say, for the coefficients of $A_1, \ldots, A_{37}$.

It is still impractical to directly check each of these 2^{37} elements to determine which are idempotent. We exploit the fact that the multiplication comes from formula (2.1). For example, suppose the product $\sigma * \rho$ contains an endomorphism of G. Then it is necessary that ρ be an endomorphism (since $\mathrm{Dom}\,\psi_x \subseteq \mathrm{Dom}\,\rho$ for each x). In our case, among the first 37 A_i's, $A_1, \ldots, A_7$ correspond to endomorphisms. So for the coefficients $a_1, \ldots, a_7$, we need only check products of the form $A_j * A_k$ with $k = 1, \ldots, 7$. Better yet, it turns out that most of these products do not contain $A_1, \ldots, A_7$; those that do correspond to $j = 1, \ldots, 13$. So our first step is to solve the resulting 7 equations in 13 unknowns. The solutions to this system will not all be idempotents, but the $a_1, \ldots, a_7$ will be correct.

In the second step, we notice that $A_1, \ldots, A_{12}$ occur only in products of the form $A_j * A_k$ with $j = 1, \ldots, 19$ and $k = 1, \ldots, 12$ ($A_8, \ldots, A_{12}$ correspond to homomorphisms $\rho : D_8 \to SD_{16}$). There result 12 equations in 19 unknowns. But we need only check the five new equations in the six new unknowns for each of the solutions from step 1. Again, the solutions will not all be idempotents, but now the $a_1, \ldots, a_{12}$ will be corect.

Continuing in this manner, all of the idempotents were found. In the actual calculations, nine steps were used (corresponding to the nine conjugacy classes of nontrivial subgroups) and about 2 million elements were checked (instead of 2^{37}). The final output gave 409,953 idempotents, so $K(G,G) \otimes Z/2$ has 819,906 idempotents.

Given the complete list of idempotents, it is easy (for a machine) to find a primitive orthogonal decomposition of the identity. First pick any idempotent e; then $1 = (e) + (1 + e)$ is an orthogonal sum. Next check all idempotents f to see if f and $e + f$ are orthogonal (this happens iff $e * f = f * e = f$). If no such f exists, then e is primitive; otherwise $1 = (f) + (e + f) + (1 + e)$ is an orthogonal sum.

Proposition 4.2

There are primitive orthogonal idempotents $e_1, \ldots, e_4$ in $K(G,G) \otimes Z/2$ with $1 = \sum e_i$.

It follows from Section 2 that a complete wedge decomposition of BG has four summands.

5. SOME HOMOLOGICAL CALCULATIONS

We still assume that $G = SD_{16}$. Since $\overline{Aut}(G) \cong Z/2$, BG has exactly one dominant summand (3.3). The other three indecomposable summands must be associated to proper subgroups and all of these are known. Let D_8 and Q_8 denote the dihedral and quaternionic groups of order 8.

Proposition 5.1

1. $BZ/2$, $BZ/4$, and $BZ/8$ are indecomposable,
2. $B(Z/2 \times Z/2) \simeq BZ/2 \vee BZ/2 \vee L(2) \vee L(2) \vee BA_4$
3. $B(D_8) \simeq BZ/2 \vee BZ/2 \vee L(2) \vee L(2) \vee BPSL_2(7)$,
4. $B(Q_8) \simeq \sum^{-1}(BS^3/BN) \vee \sum^{-1}(BS^3/BN) \vee BSL_2(3)$,

where $L(2) = \sum^{-2}(Sp^4S^0/Sp^2S^0)$, A_4 is the alternating group on four letters, and $N \subseteq S^3$ is the normalizer of a maximal torus. $BZ/2$ is associated to $Z/2$, $L(2)$ is associated to $Z/2 \times Z/2$, and all other summands are dominant.

Proof. The decompositions were found in [6]. Most of the summands can be shown to be indecomposable using (3.2), induction, and Poincaré series. For BA_4 and $BPSL_2(7)$ one also uses the action of the Steenrod algebra (to show that $L(2)$ cannot split off).

To determine which of these summands occur in BG, we use the explicit idempotents from Section 4 to calculate Poincaré series. A homomorphism $\rho : H \to G$ gives a map $tr^* \circ \rho^* : H^*(BG) \to H^*(BH) \to H^*(BG)$ on cohomology. So to understand the map $K(G,G) \otimes Z/2 \to End(H^*BG)$, we need to calculate induced homomorphisms and transfer homomorphisms.

Most of the induced homomorphisms are easy to calculate after the restriction maps to subgroups have been found. We list some of the cohomologies and restriction maps. We also give the action of the Steenrod algebra, since this will be useful in the transfer calculations below.

Let D_8 and Q_8 be the subgroups of SD_{16} generated by $\{a^2,b\}$ and $\{a^2,ab\}$, respectively, and let $\phi : D_8 \to SD_{16}$ and $\psi : Q_8 \to SD_{16}$ be the inclusions.

Proposition 5.2

1. $H^*(SD_{16}) = Z/2[x,y,u,P]/(x^2 + xy, x^3, xu, u^2 + x^2P + y^2P)$, with $|x| = |y| = 1$, $|u| = 3$, $|P| = 4$. $Sq^1(u) = Sq^1(P) = Sq^3(P) = 0$, $Sq^2(u) = y^2u + xP + yP$, and $Sq^2(P) = x^2P + y^2P$.
2. $H^*(D_8) = Z/2[\bar{x},\bar{y},\bar{w}]/(\bar{x}^2 + \bar{x}\bar{y})$, with $|\bar{x}| = |\bar{y}| = 1$, $|\bar{w}| = 2$. $Sq^1(\bar{w}) = \bar{y}\bar{w}$.
3. $H^*(Q_8) = Z/2[\tilde{x},\tilde{y},\tilde{P}]/(\tilde{x}^2 + \tilde{x}\tilde{y} + \tilde{y}^2, \tilde{x}^2\tilde{y} + \tilde{x}\tilde{y}^2)$, with $|\tilde{x}| = |\tilde{y}| = 1$, $|\tilde{P}| = 4$. $Sq^1(\tilde{P}) = Sq^2(\tilde{P}) = Sq^3(\tilde{P}) = 0$.
4. $\phi^*(x) = 0$, $\phi^*(y) = \bar{y}$, $\phi^*(u) = \bar{y}\bar{w}$, and $\phi^*(P) = \bar{w}^2$.
5. $\psi^*(x) = \psi^*(y) = \tilde{y}$, $\psi^*(u) = \tilde{x}^2\tilde{y}$, and $\psi^*(P) = \tilde{P}$.

<u>Proof</u>. The cohomologies are calculated in [1,6,8]. The Steenrod operations for D_8 and Q_8 are also given. The restriction homomorphisms were partially computed in [1]. By comparing their results with the possible actions of the Steenrod algebra, the proof can be completed.

As an example, we calculate $Sq^1(u)$, $\phi^*(u)$, and $\psi^*(u)$. $H^3(Q_8)$ is one-dimensional and $\psi^*(u) \neq 0$ by the Gysin sequence (5.3) below, so $\psi^*(u) = \tilde{x}^2\tilde{y}$. From [1] we have $\phi^*(u) = \bar{y}\bar{w} + c\bar{x}^3$ for some $c \in Z/2$.

Let $Sq^1(u) = \alpha y^4 + \beta yu + \gamma P$. Then $\gamma\tilde{P} = \psi^*Sq^1(u) = Sq^1\psi^*(u) = Sq^1(\tilde{x}^2\tilde{y}) = 0$, so $\gamma = 0$. Also $\alpha\bar{y}^4 + \beta\bar{y}^2\bar{w} + \beta c\bar{x}^4 = \phi^*(\alpha y^4 + \beta yu) = \phi^*Sq^1(u) = Sq^1\phi^*(u) = Sq^1(\bar{y}\bar{w} + c\bar{x}^3) = c\bar{x}^4$ so $\alpha = \beta = c = 0$.

To calculate the transfer homomorphisms we use the following two propositions. The first is a form of the Gysin sequence and the second is a special case of the double coset formula. Let J be a subgroup of index two in the 2-group K, and let $i : J \to K$ be the inclusion. The homomorphism $\beta : K \to K/J \cong Z/2$ induces an element $\beta \in H^1(BK)$.

Proposition 5.3 [5, Corollary 12.3]

The sequence is exact:

$$\cdots \xrightarrow{tr^*} H^{i-1}(BK) \xrightarrow{\cup\beta} H^i(BK) \xrightarrow{i^*} H^i(BJ) \xrightarrow{tr^*} \cdots$$

Proposition 5.4

Let $k \in K - J$. Then the composition $H^*(BJ) \xrightarrow{tr^*} H^*(BK) \xrightarrow{i^*} H^*(BJ)$ equals $id^* + c_k^*$, where c_k denotes conjugation by k.

Propositions 5.2, 5.3, and 5.4 give enough information to calculate the cohomology images of the explicit idempotents from Section 4. The Poincaré series of the three nondominant summands can then be found and the summands identified using Proposition 5.1.

Proposition 5.5

$BSD_{16} \simeq BZ/2 \vee \Sigma^{-1}(BS^3/BN) \vee L(2) \vee X$, where X is a dominant summand. The Poincaré series of X is $(t^3 + t^4 + t^5 - t^7)/[(1 - t^3)(1 - t^4)]$.

ACKNOWLEDGMENT

We would like to thank Ian Hambleton for suggesting Proposition 5.3 as an aid for calculating the transfer homomorphism and the Mathematics Department at the University of Minnesota, Duluth for the use of their facilities during a recent visit.

REFERENCES

1. L. Evens and S. Priddy, The cohomology of the semi-dihedral group, Contemp. Math., 37, (1985), 61–72.
2. J. C. Harris, Stable Splittings of Classifying Spaces, Thesis, University of Chicago (1985).
3. J. C. Harris and N. J. Kuhn, Stable decompositions of classifying spaces of abelian p-groups, Math. Proc. Cambridge Philos. Soc., 103, (1988), 427–449.
4. J. P. May, Stable maps between classifying spaces, Contemp. Math., 37, (1985), 121–130.
5. J. W. Milnor and J. D. Stasheff, Characteristic Classes, Annals of Math. Studies no. 76, Princeton, University Press (1974).
6. S. A. Mitchell and S. B. Priddy, Symmetric product spectra and splittings of classifying spaces, Am. J. Math., 106, (1984), 219–232.

7. G. Nishida, Stable homotopy type of classifying spaces of finite groups, preprint (1985).
8. D. Quillen, The mod-2 cohomology rings of extra-special 2-groups and the spinor groups, Math. Ann., 194, (1971), 197–212.

10

Algorithms for Computing the Cohomology of Nilpotent Groups

LARRY A. LAMBE*

North Carolina State University
Raleigh, North Carolina

1. INTRODUCTION

The purpose of this expository note is to show how computers can be used to calculate the cohomology of certain groups. These examples give a good introduction to the situations encountered in a class of iterated fiber bundles [14]. The classifying spaces of the groups involved in this chapter are such bundles. We have used computer calculations successfully to compute new examples which have generated new mathematical conjectures [12]. Our intention is to give an idea of what is involved in this process, indicating how a complete solution to the problem of calculating the cohomology of torsion-free nilpotent groups may be programmed. Two sets of algorithms are available, depending on whether answers are required over the integers Z or the rational numbers Q. If we are allowed to leave Z and go to subrings of Q, then we can make calculations using a program [6] which computes the cohomology of Lie algebras and records generators in each degree and feed the output to a REDUCE program, which generates the actual cochains corresponding to these generators. In the case that we must remain over Z, things are more complicated. The Lie algebra cohomology program produces output which is processed symbolically to produce an encoding of a certain differential graded algebra, which is

*Current affiliation: University of Illinois at Chicago, Chicago, Illinois

to be fed into another program which calculates the cohomology of such objects and records generators in each degree, and then the process is repeated. We will, in fact, show that it is possible to program the entire Eilenberg-Moore spectral sequence from E_2 to E_∞ for a large class of objects. The actual group cohomology cochains can, once again, be read from this information, but we will leave that for another note. It is noted in Section 3 that the calculation of such cochains in degree 2 leads to an enumeration of nilpotent groups. In the overall picture, a phenomenon well known in homotopy theory is illustrated, namely that the descent from the rational category to the integers involves a considerable increase in complexity. In Section 3 we assume some familiarity with REDUCE but nothing beyond what a quick glance at the manuals [9,16] will give. In Section 4 we assume some familiarity with LISP but again nothing beyond what can be found in the first few chapters of most beginning LISP texts. Most of the mathematical background can be obtained from [7,13,15].

2. GROUP COHOMOLOGY BACKGROUND AND THE CASE n = 0 OF AN INDUCTION

Consider the extension cocycle [15, pp. 111, 137] associated to a central extension of groups

$$0 \to A \overset{\pi}{\to} E \to G \to 1 \qquad [2.1]$$

By this we mean that A is contained in the center of E and the sequence (2.1) is exact. We write A additively. Since π is onto, we can choose a function $s: G \to E$ such that, for all $x \in G$, we have $\pi s(x) = x$. Generally, the function s is not a homomorphism. The obstruction

$$f(x,y) = s(x)s(y)s(xy)^{-1}$$

yields a well-defined function $f: G \times G \to A$ which satisfies the cocycle condition:

$$f(y,z) - f(xy,z) + f(x,yz) - f(x,y) = 0 \qquad \text{for all } x,y,z \in G$$

coming from the associativity of multiplication. We obtain a group isomorphism, $G \times A \cong E$, by giving the set $G \times A$ the operation

$$(g,z)(g',z') = (gg', z + z' + f(g,g'))$$

In one direction, this isomorphism takes (g,z) to $s(g)z$. Write the constructed group as G_f. Another extension cocycle h will give rise to another group G_h, and a vertical isomorphism exists making the diagram

$$\begin{array}{ccccccccc} 0 & \to & A & \to & G_f & \to & G & \to & 1 \\ & & \downarrow & & \downarrow & & \downarrow & & \\ 0 & \to & A & \to & G_h & \to & G & \to & 1 \end{array} \qquad [2.2]$$

commute if and only if there is a $k : G \to A$ such that for all $x,y \in G$

$$f(x,y) - h(x,y) = k(y) - k(xy) + k(x)$$

All of this may be summarized succinctly by defining the abelian groups

$$C^k(G;A) = \{f \mid f : G^k \to A\} \qquad k = 1,2,3 \qquad [2.3]$$

and defining a sequence of operators:

$$C^1 \overset{\delta}{\to} C^2 \overset{\delta}{\to} C^3 \qquad [2.4]$$

by

$$\delta(f)(x,y) = f(y) - f(xy) + f(x)$$

$$\delta(f)(x,y,z) = f(y,z) - f(xy,z) + f(x,yz) - f(x,y)$$

Noting that $\delta\delta = 0$, we can form the quotient group $\ker(\delta)/\mathrm{im}(\delta)$, and the isomorphism classes of E relative to the diagram (2.2) are in bijective correspondence with this group:

$$\ker(\delta)/\mathrm{im}(\delta) = H^2(G;A) \qquad [15, \text{ p. } 112]$$

Baer showed that for G free abelian we have an isomorphism:

$$H^2(G;A) = \mathrm{Hom}(\Lambda^2(G),A) \qquad [2.5]$$

where $\Lambda^2(G)$ denotes the exterior product of G with itself. In particular, if we take $A = Z$, we get

$$\begin{aligned} H^2(Z^n,Z) &= \mathrm{Hom}(\Lambda^2 Z^n,Z) \\ &= \Lambda^2(Z^n)^* \end{aligned} \qquad [2.6]$$

where $(Z^n)^* = \mathrm{Hom}(Z^n, Z)$ is freely generated by the elements $e^i \wedge e^j$ for $1 \leq i < j \leq n$, and $\{e^1, \ldots, e^n\}$ is the standard dual basis of $(Z^n)^*$. Thus, given integers, $1 \leq i < j \leq n$, we have the group $Z^n \times_f Z$ corresponding to the element $f = \sum f_{ij} e^i \wedge e^j$. There are some interesting groups in this class. For example, the 3×3 upper unitriangular matrices with 1's along the diagonal are given (up to isomorphism) by the case $n = 2$ and $f = e^1 \wedge e^2$.

The definition of the differentials in (2.4) may be given uniformly for all k by

$$\begin{aligned}\delta(f)(x_1, \ldots, x_{k+1}) = {} & f(x_2, \ldots, x_{k+1}) \\ & + \sum (-1)^i f(x_1, \ldots, x_i x_{i+1}, \ldots, x_{k+1}) \\ & + (-1)^{k+1} f(x_1, \ldots, x_k)\end{aligned} \qquad [2.7]$$

We obtain a graded abelian group $\sum H^k(G;A) = H^*(G;A)$ by taking the kernel mod the image at each stage. More is true. If R is a commutative ring with 1, we get a graded R-algebra structure on $H^*(G;R)$ as follows: Given functions $f : G^r \to R$, $g : G^s \to R$, take their product (cup product) to be the function

$$f \cup g : G^{r+s} \to R \qquad [2.8]$$

defined by

$$(f \cup g)(x_1, \ldots, x_{r+s}) = f(x_1, \ldots, x_r) g(x_{r+1}, \ldots, x_{r+s}) \qquad [2]$$

Cup product of functions is obviously associative and noncommutative in general. It does pass to cohomology to give an associative graded commutative product. The graded commutativity can be established by the existence of a collection of functions (cup-1 product):

$$\cup_1 : C^r \otimes C^s \to C^{r+s-1} \qquad [2.9]$$

which satisfy

$$\delta(f \cup_1 g) = f \cup g - (-1)^{|f||g|} g \cup f - \delta(f) \cup_1 g - (-1)^{|f|} f \cup_1 \delta(g)$$

$$[7] \qquad [2.10]$$

With these definitions we obtain an isomorphism [2],

$$H^*(Z^n;Z) \to \Lambda^*(Z^n)^* \cong E[e_1,\dots,e_n] \quad \text{generalizing (2.6)} \qquad [2.11]$$

that is, the cohomology of the free abelian group on n generators is the exterior algebra on n generators of dimension 1.

3. FINITELY GENERATED NILPOTENT GROUPS OVER SUBRINGS OF Q

In general, we may start with a given group G, choose $f \in H^2(G;R)$, and form G_f. A choice of $k \in H^2(G_f;R)$ then gives another group and so on. We can consider the class of groups obtainable from a given group G in this way. A well-known result in group theory (although it is not always stated this way) says that if we start out with $G = Z/p$ and $R = Z/p$, then we get the class of finite p-groups. If we begin with $G = Z$ and $R = Z$, then we obtain the class of finitely generated torsion-free nilpotent groups $\mathfrak{M}$. It is the class $\mathfrak{M}$ that we will deal with here. We begin with an observation of P. Hall about the group law of such groups.

Proposition 3.1

For $G \in \mathfrak{M}$, we have an isomorphism $G \cong (Z^n, \rho)$ where the group operation ρ is a polynomial function

$$\rho : Z^n \times Z^n \to Z^n \qquad [8]$$

Remark 3.2. A consequence of (3.1) is that we may choose our defining cocycles for the class $\mathfrak{M}$ to be polynomial functions. We will see this below. In general, such functions are given by polynomials with rational coefficients which take integers to integers. These are all given by linear combinations (with integer coefficients) of monomials of the form

$$\binom{n_1}{r_1}\cdots\binom{n_k}{r_k}$$

where the binomial coefficient $\binom{n}{r}$ is given by the polynomial in n,

$$\binom{n}{r} = \frac{n(n-1)\cdots(n-(r+1))}{r!} \qquad \text{for a fixed integer } r \quad [17]$$

We will describe an algorithm [13] which gives an effective procedure for constructing group cohomology cochains from Lie algebra cocycles. In fact, it will be clear that the algorithm can easily be implemented in a symbolic manipulation language such as REDUCE for experimentation. Using the representation of G as (R^n, ρ), for $R = Z[1/2, 1/3, \ldots, 1/\rho] \subset Q$, the smallest subring of the rationals which contains all the coefficients of b, we may form the Lie algebra $\mathcal{L}(G) = (R^n, [\ ,\ ])$ where

$$[x,y] = b(x,y) - b(y,x)$$

Here b is the homogeneous degree 2 term of the group law:

$$\rho(x,y) = x + y + b(x,y) + O(\geq 3)$$

We can form the "exterior" complex of $\mathcal{L}(G)$ over R [1,4,11] as follows. Let $\wedge^r$ = rth exterior power of $\mathcal{L}(G)^* = \mathrm{Hom}(\mathcal{L}(G), R)$ and let $\{e^i\}$ be a basis for $\mathcal{L}(G)^*$ dual to a basis $\{e_i\}$ for $\mathcal{L}(G)$.

$$\wedge^1 \xrightarrow{d} \wedge^2 \xrightarrow{d} \cdots \xrightarrow{d} \wedge^n$$

is given by extending

$$d(e^i) = \sum C^i_{jk} e^j \wedge e^k \qquad [3.3]$$

as a derivation using the rule

$$d(\alpha \wedge \beta) = d(\alpha) \wedge \beta + (-1)^{|\alpha|} \alpha \wedge d\beta$$

where the C^i_{jk} are the structure constants of the Lie algebra:

$$[e_i, e_j] = \sum C^k_{ij} e_k$$

3.4 Right Invariant Vector Fields and Forms

Consider the real Lie group (R^n, ρ) obtained from the group law ρ by allowing real entries. We obtain a basis E_j for the right invariant vector fields on (R^n, ρ) by taking

$$E_j = \sum a_{ij} \frac{\partial}{\partial x_i} \qquad [3.5]$$

where

$$a_{ij}(x) = \frac{\partial}{\partial y^j} |_0 \rho_i(y,x)$$

A (dual) basis for the right invariant 1-forms is given by

$$e^j = \sum \omega_{ij}\, dx^i \qquad [3.6]$$

where

$$\omega_{ij}(x) = \frac{\partial}{\partial y^i} |_0 \rho_j(y,x^{-1})$$

If $A = (a_{ij})$, $\Omega = (\omega_{ij})$, then $\Omega = (A^{-1})^T$ [13]. The right invariant vector fields and forms can be generated in REDUCE by the simple code in Table 1, where an array p of length n has been defined and contains the components ρ_i of the group law ρ and the right invariant vector fields E_j above are computed and recorded in the matrix vf. The forms ω_{ij} are recorded in the matrix fms. The output of the program in Table 1 is given in Table 2.

3.7 A Family of Simplices

Let $\Delta_k = \text{con}\{e_0,\ldots e_k\}$ be the convex hull of the set of vectors $\{e_0,e_1,\ldots,e_k\}$ where $e_0 = 0$ and $e_1,\ldots,e_k$ are the standard basis vectors. For k elements $d_1,\ldots,d_k$ $\quad (Z^n,\rho)$ define a simplex $s(d_1,\ldots,d_k)$ in $R^n \in$ by extending the correspondence

$$\Delta_k \to R^n$$

given by

$$e_0 \longmapsto 0$$

$$e_1 \longmapsto d_1$$

$$\cdots$$

$$\cdots$$

$$\cdots$$

$$e_k \longmapsto d_k d_{k-1} \cdots d_2 d_1$$

linearly. It is straightforward to show that the correspondence $\mathcal{G}$ given by

Table 1

```
%
%
% Generate right invariant forms corresponding to elements
% in the Koszul complex of the Lie algebra associated to a
% nilpotent group using E. Schruefer´s EXCALC package.
%
%
% EXCALC must be loaded for this example.
off echo;

% aux function for the example:

   procedure binco(n,k);
     for i := 1:k product (n - i + 1)/i;

%

n:=4$
array p(n);
operator x,y;
%
for i := 1:2 do
    p(i) := y(i) + x(i);
p(3) := y(3) + x(3) - y(2) * x(1);
p(4) := y(4) + x(4) - y(3) * x(1) + y(2) * binco(x(1),2);
%
% form the matrix of coefficients of the right invariant
% vector fields
matrix vf(n,n);
for i:=1:n do
    for j:=1:n do
        if j > i then vf(i,j):=0
          else if i > j then vf(i,j):=sub(y(j)=0,df(p(i),y(j)))
               else vf(i,j):=1;
%
matrix fms;
fms := tp(vf ** (-1))$
%
% declare the dimension of the underlying space
spacedim n;
%
% form the invariant 1-forms, i is an arbitrary index = 1 .. n,
% dx(i),ee(i) are 1 - forms.
pform dx(i)=1,ee(i)=1;
```

Table 1 (Continued)

```
%
for j:=1:n do
    ee(j) := for i:=1:n sum fms(i,j) * dx(i);
%
% write out ee1,...,ee4:
write "The 1-forms: ";
for i:=1:4 do write "E(",i,") = ", ee(i);
%
% write out the two forms:
write "The 2-forms: ";
for i:=1:3 do
    for j:=(i+1):4 do
    write "E(",i,")∧E(",j,") = ", ee(i)∧ee(j);
%
% write out the non-zero 3-forms:
write "The non-zero 3-forms: ";
for i:=1:2 do
    for j:=(i+1):3 do
        for k:=(j+1):4 do
            << treform:=ee(i)∧ee(j)∧ee(k);
               if treform neq 0 then
                 write "E(",i,")∧E(",j,")∧E(",k,") = ", treform
            >>

%
% end example.
on echo;
;end;
```

Table 2

```
reduce
SLISP : 3772512 BYTES



    REDUCE 3.2, 15-Apr-85 ...

    1:
load excalc;

    2:
in forms;
    %
    %
    % Generate right invariant forms corresponding to elements
    % in the Koszul complex of the Lie algebra associated to a
    % nilpotent group using E. Schruefer's EXCALC package.
    %
    %
    % EXCALC must be loaded for this example.
    off echo;


    BINCO

    P(3) := X(3) - X(1)*Y(2) + Y(3)

                                 2
    P(4) := (2*X(4) + X(1) *Y(2) - 2*X(1)*Y(3) - X(1)*Y(2) + 2*Y(4))/2

    The 1-forms:

              1
    E(1) = DX

              2
    E(2) = DX

                     2     3
    E(3) = X(1)*DX  + DX

                  2    2          2            3      4
    E(4) = (X(1) *DX  + X(1)*DX  + 2*X(1)*DX  + 2*DX )/2
```

Table 2 (Continued)

```
The 2-forms:

                  1    2
E(1)∧E(2) = DX ∧DX

                       1    2      1    3
E(1)∧E(3) = X(1)*DX ∧DX  + DX ∧DX

                    2     1    2          1    2            1    3
E(1)∧E(4) = (X(1) *DX ∧DX  + X(1)*DX ∧DX  + 2*X(1)*DX ∧DX  + 2*

              1    4
             DX ∧DX )/2

                  2    3
E(2)∧E(3) = DX ∧DX

                       2    3      2    4
E(2)∧E(4) = X(1)*DX ∧DX  + DX ∧DX

                    2     2    3          2    3            2    4
E(3)∧E(4) = (X(1) *DX ∧DX  - X(1)*DX ∧DX  + 2*X(1)*DX ∧DX  + 2*

              3    4
             DX ∧DX )/2

The non-zero 3-forms:

                       1    2    3
E(1)∧E(2)∧E(3) = DX ∧DX ∧DX

                            1    2    3      1    2    4
E(1)∧E(2)∧E(4) = X(1)*DX ∧DX ∧DX  + DX ∧DX ∧DX

                         2     1    2    3          1    2    3
E(1)∧E(3)∧E(4) = (X(1) *DX ∧DX ∧DX  - X(1)*DX ∧DX ∧DX  + 2*X(1)*

                   1    2    4        1    3    4
                  DX ∧DX ∧DX  + 2*DX ∧DX ∧DX )/2

                       2    3    4
E(2)∧E(3)∧E(4) = DX ∧DX ∧DX

;

end;

3:
bye;
*** END OF RUN
```

$$\mathscr{I}: \Lambda^k \to C^k(G;R) \qquad [3.8]$$

where

$$\mathscr{I}(\alpha)(d_1,\ldots,d_k) = \int_{s(d_1,\ldots,d_k)} \alpha$$

for $R \subset Q$ is a cochain map from the exterior complex to the functional cochain complex defined in (2.3), extended to all k in (2.7). Under certain restrictions on R in [13], and under less restrictive conditions in [3], it follows that (2.8) induces an isomorphism in cohomology. We can, once again, easily compute the cochain map $\mathscr{I}$ in REDUCE using the integration operator int and the substitution operator sub to calculate the integral

$$\int_{s(d_1,\ldots,d_k)} \alpha = \int_{\Delta_k} s(d_1,\ldots,d_k)^*(\alpha)$$

Notice that the result is a polynomial in the variables $d_1,\ldots,d_k$. In summary, if we have an effective means to compute the cohomology of the exterior complex (3.3), then we have an inductive enumeration of the class $\mathfrak{M}$ (suitably localized). A set of programs to efficiently record this exterior complex over the integers and to compute its cohomology and record a set of generators in each dimension has been written by Len Evens [6]. These programs have been used in [12] in connection with the problem mentioned in the sections below. There we consider the problem of obtaining all the torsion information over the integers. The algorithms necessary are considerably more complex and are really just the first stage in a sequence of increasing complexity illustrated by the theory in [14].

4. INTEGER COEFFICIENTS

There are examples of finitely generated torsion-free nilpotent groups which have torsion in their cohomology which is not detected by the methods above. The 4-generated two-step free nilpotent group is such an example. This and other examples are discussed in [12]. We will review the algorithms involved and give an implementation in LISP. We would like to remark that most of what we know about using LISP for symbolic manipulation was learned from reading the source code for REDUCE which is included in the standard releases of that language. Everything presented here for the case of two-step groups has been generalized in joint work with J. D. Stasheff [14] which is applicable to n-step nilpotent groups. The theory for the two-step cases presented here appears in [12], which is in turn an application of the material in [7]. An independent exposition of the theory for the two-step case in arbitrary characteristic appears in [10].

4.1 The Theorem of Eilenberg and Moore and Extra Terms in the Differential of the Exterior Complex

We will describe a complex due to V. K. A. M. Gugenheim and J. P. May [7] which, in particular, can be used to compute the integral cohomology of two-step nilpotent groups. The starting point is the theorem of Eilenberg and Moore [5] for the fiber square

$$\begin{array}{ccccc} T^f & \to & BD & \to & ET_f \\ & & \downarrow & & \downarrow \\ & & T^b & \to & T^f \end{array} \qquad [4.2]$$

which is obtained from the central extension (2.1) for $Z^f = A$, $Z^b = G$, and $D = E$. By Baer's result (2.5), the extension cocycle can be represented as an f-tuple of integer forms $(\phi_1, \ldots, \phi_f)$. By using the representability of $H^2(\_;A)$ in the homotopy category, the ϕ_i determine a continuous map $\Phi : T^b \to BT^f$ and the classifying space of the group D is equivalent to the pull-back of the universal bundle over BT^f by this map [12]. The reason that we place ourselves in this geometric setting is so that we can quote the following:

Theorem 4.3 (Eilenberg-Moore)

Consider the fiber square (4.2). There is an isomorphism of cohomology rings

$$H^*(BD;Z) \cong \mathrm{Tor}_{C^*(K)}(C^*(T^b;Z),Z)$$

where $C^*(;Z)$ denotes the cochain algebra of singular cochains with Z coefficients; $K = C^*(BT^f;Z)$, $C^*(T^b;Z)$, and $C^*(ET^f;Z) \cong Z$ are differential graded modules over the differential graded algebra $C^*(BT^f;Z)$ via the natural maps obtained from the fiber square; and Tor denotes differential Tor.

Now the point is to cut the standard complex for this differential Tor down to a manageable size. This can be done in such a way that the underlying Z-module structure of the complex is the same as that of the exterior complex for the corresponding Lie algebra (3.3). The differential in the exterior complex, however, needs alteration. Interestingly, the correct thing to do turns out to be to add on some extra terms. That is, we have the complex

$$\Lambda^1 \xrightarrow{d} \Lambda^2 \xrightarrow{d} \cdots \xrightarrow{d} \Lambda^n \qquad [4.4]$$

which is additively the same as the Z-module structure for the exterior algebra $E[u_1,\ldots,u_f,p_1,\ldots,p_b]$, with a differential d given by the sum

$$d = d_1 + d_2 + d_3 + \cdots$$

where d_1 is the exterior differential for the Lie algebra $(Z^{f+b},[\ ,\ ])$ corresponding to the group $D = (Z^{f+b},\rho)$. While d_1 is a derivation of the exterior algebra, the terms d_i, $i \geq 2$, and the sum $d = \sum d_i$ are, in general, not. To describe this differential, consider the natural bigrading of the complex $X = E[u_1,\ldots,u_f,p_1,\ldots,p_b]$ given by decomposing it as the tensor product $E[u_1,\ldots,u_f] \otimes E[p_1,\ldots,p_b]$. Thus define α to have bidegree (r,s) if $\alpha = \alpha_1 \otimes \alpha_2$ where α_1 is an r-form in the u_i and α_2 is an s-form in the p_j. Write the component of X of bidegree (r,s) as $X_{r,s}$. The differential d satisfies the following:

1. $d_k : X_{r,s} \to X_{r-k,s+k-1}$.
2. $\sum_{i+j=r} d_i d_j = 0$ for each r.
3. $d_k(up) = d_k(u)p$, $u \in E[u_1,\ldots,u_f]$, $p \in E[p_1,\ldots,p_b]$. [4.5]
4. If we filter X by $F_k(X) = \sum X_{m,*}$, $k \geq 0$, then d induces a spectral sequence, $\{E_k,d_k\}$.
5. $d_k(u_i) = \sum \tau(i,j)u_{i-j} \otimes g_j$.

where, generally,

$$g_j = g((\ldots(\phi_{j_1} \cup_1 \phi_{j_2} \cup_1 \cdots) \cup_1 \phi_{j_t})$$

for

$$j = \{j_1, \ldots, j_t\}$$

The function g is a multiplicative map which is described below, i is an increasing subset of $\{1, \ldots, f\}$, the sum ranges over all increasing subsets $j \subset i$, $i - j$ is the complement of j in i, and for $j = \{j_1, \ldots, j_t\} \subset i$ the sign τ is given by

$$\tau(i,j) = (-1)^{|i-j| + \varepsilon(j)}$$

where

$$\varepsilon(j) = \sum_{k=1}^{t} j_k - k$$

The operation $\cup_1$ in (4.5.5) is the cup-1 product mentioned in (2.9). It is completely determined on the cochains involved in (4.5.5) by the following facts:

$$(\alpha \cup \beta) \cup_1 \gamma = (-1)^{|\alpha|} \alpha \cup (\beta \cup_1 \gamma)$$
$$+ (-1)^{|\beta||\gamma|} (\alpha \cup_1 \gamma) \cup \beta \qquad \text{(Hirsch formula)} \qquad [4.6.1]$$

and

$$C^1(G;Z) \otimes C^k(G;Z) \to C^k(G;Z)$$

is given by

$$(\alpha \cup_1 \beta)(g_1, \ldots, g_k) = \alpha(g_1 \cdots g_k) \beta(g_1, \ldots, g_k) \qquad [4.6.2]$$

From the Hirsch formula, which just says that $\_\cup_1 \gamma$ is a derivation of (the noncommutative) cup product, we have for one-dimensional cochains $f_{i_1}, \ldots, f_{i_n}$ and any cochain β,

$$(f_{i_1} \cup \cdots \cup f_{i_n}) \cup_1 \beta = \qquad [4.6.3]$$

$$\sum (-1)^{|([n-t]|\beta|+t-1)|} f_{i_1} \cup \cdots \cup f_{i_{(t-1)}} \cup (f_{i_t} \cup_1 \beta) \cup f_{i_{(t+1)}} \cup \cdots \cup f_{i_n}$$

Also, (4.5.5) shows that only two-dimensional cochains show up on the right of any $\cup_1$ so we need to consider next the cup-1 product of a 1-cochain and a 2-cochain. First of all, the only 1-cochains that we need to consider are functions of the form P_I where $P_I : Z^b \to Z$ is a pointwise product of the projections $p_{i_1}, \ldots, p_{i_j}$ for an index set $I = \{i_1, \ldots, i_j\}$. This follows by induction. The cup-1 product of such a function with a 2-cochain $\beta = p_r \cup p_s$ is easily computed by (4.6.2). We have

$$P_i \cup_1 \beta = \sum P_{J \amalg \{r\}} \cup P_{K \amalg \{s\}} \qquad [4.6.4]$$

where the sum is over all index sets J,K such that $J \amalg K = \{i_1, \ldots, i_j\}$ and $\amalg$ denotes disjoint union.

4.7 Data Structures for the Cup-1 Product in LISP

We define a monomial to be an integer times a product of one-dimensional cochains as described above. A sum of monomials $L_1, L_2, \ldots, L_k$ will be represented by a list $L = (L_1 L_2 \cdots L_k)$. Each factor of a monomial will thus be of the form P_I where $I = \{i_1, \ldots, i_t\}$ is some index set. If a monomial has the form $nP_{I_1}P_{I_2} \cdots P_{I_m}$ then we will represent this by the dotted pair (I.n) where $I = (I_1 \ldots I_m)$ is simply the list of the indices. Thus, for example, the cochain

$$-P_{ik}P_qP_w - 3P_kP_{ij}P_v + P_iP_wP_{qr}$$

is represented as the list $(L_1\ L_2\ L_3)$ where, for example, L_2 is the list

(((k) (ij) (v)). −3)

To illustrate the ideas we will use the notation

(function arg1 arg2 ...) : LISP expression

to give the definition of a function. The definitions will be somewhat abbreviated; for example, we leave out initial conditions. The first function we need to deal with is $(\cup_1\ L\ \beta)$. Since $\cup_1$ is bilinear, β distributes over the list L and hence we make the definition

```
(∪1 L β) : (append
              (∪1b (car L) β)
              (∪1 (cdr L) β))                              [4.7.1]
```

Where $\cup_1$ b is a function designed to deal with the $\cup_1$ product of a monomial and a cochain β. The Hirsch formula applied to the case $(x \cup y) \cup_1 \beta$ where $|x| = 1$ and $|\beta| = 2$ gives:

$$(x \cup y) \cup_1 \beta = (x \cup_1 \beta) \cup y - x \cup (y \cup_1 \beta)$$

so that we need a left cup product and a right cup product, i.e., we need (rtcup lst b) and (lftcup c lst) where b and c are monomials, in order to deal with the function ($\cup_1$b (car L) β) above. Here (car L) is a list that looks like (L_1.n) where n is an integer and L_1 is a list of lists of the form $(i_1 \cdots i_k)$ where the i_j are integers. Thus rtcup should just append its second argument to the end of each element of the list L_1, which should be its first argument. We will deal with the coefficient n later. Thus we define

```
(rtcp lst b) : (mapcar
                 '(lambda (z) (append z b))
                 lst)                                      [4.7.2]
```

The function lftcup is more complicated even though its first argument is one-dimensional because its second argument is a linear combination. It should append its first argument to the front of the car of each element of the list which is its second argument. Thus, the definition of lftcup is the same as that of rtcup with the function append replaced by carappend where carappend has the definition

```
(carappend u w) : (cons
                     (append
                        (list u) (car w))
                        (cdr w))                           [4.7.3]
```

Now we need to deal with the signs. The definition of $\cup_1$ b is (keep in mind that its first argument looks like (L_1.n) above)

```
(∪₁ b M β) : (append
                 (mksgn (cdr M)
                   (rtcup
                     (∪₁ a (caar M) β))
                     (cdar M)))
                 (sgn (minus (cdr M))
                   (lftcup
                     (caar M)
                     (∪₁ b (cons (cdar M) 1) β))))          [4.7.4]
```

We use the auxiliary functions mksgn and sgn, whose definitions are

```
(mksgn r lst) : (mapcar
                   '(lambda (z) (cons z r))
                   lst)                                    [4.7.5]
(sgn r lst) : (mapcar
                 '(lambda (z) (cdrmul r z))
                 lst)                                      [4.7.6]
```

where crdmul is

```
(cdrmul r w) : (cons
                  (car w)
                  (times (cdr w) r))                       [4.7.7]
```

Finally, we come to $\cup_1 a$ which corresponds to (4.6.4). Its first argument is a list $I = (i_1, \ldots, i_j)$ and its second argument is β, which we now assume to be given in the form of a list $\beta = (r\ s)$. We have

```
(∪₁ a I β) : (mapcar
                '(lambda (z) (precup1 z I β))
                (powset (length I)))                       [4.7.8]
```

where (powset n) is the list of all subsets of 1,2,...,n each listed in its natural order and the function precup1 is defined by

```
(precup1 z I β) : (list
                    (append
                      (pick z I)
                      (list (car β)))
                    (append
                      (copick z I)
                      (cdr β)))                    [4.7.9]
```

Here, for an increasing set of integers z each of whose entries is $\leq$ the cardinality of z and a set I whose cardinality is at least that of z, (pick z I) is the subset of I consisting of all those elements which are in the positions given by the elements of z and (copick z I) is the complement of this set. Since pick, copick, and powset are more or less standard LISP utility functions, we omit their definitions. To illustrate the raw data generated by these functions, we will list the output of ($\cup_1$ '(i j) '(k l)) from an IBM/XT implementation:

```
((((i k) (l) (j)) . 1)
 (((k) (i l) (j)) . 1)
 (((i) (j k) (l)) . -1)
 (((i) (k) (j l)) . -1))
```

This data could easily be converted into an expression that looks more like the formula

$$\begin{aligned}(P_i \cup P_j) \cup_1 (P_k \cup P_l) = {} & P_{ik} \cup P_l \cup P_j \\ & + P_k \cup P_{il} \cup P_j \\ & - P_i \cup P_{jk} \cup P_l \\ & - P_i \cup P_k \cup P_{jl}\end{aligned}$$

that it represents, but our point is not to look at pieces of the differential but to process it in a way that will be useful in calculations; we still have to apply the function g. This map is described in detail in [12]. We will just record a formula for it here. For $\gamma \in C^k(Z^b;Z)$ the coefficient of $P_{i_1}P_{i_2}\cdots P_{i_k}$ in the linear combination for $g(\gamma)$ is given by

$$\sum \mathrm{sgn}(\sigma)\gamma(t_{\sigma(i_1)},\ldots,t_{\sigma(i_k)})$$

where the sum ranges over all permutations of the set $\{i_1,\ldots,i_k\}$. This expression is just a sum of products of Kronecker delta functions when γ is one of the cochains output above, and it is clear that this and the complete differential (4.5.5) can be encoded and the program mentioned in the introduction can be carried out. We mention here that in every case in which this has been done, including the cases of the 2-, 3-, 4-, and 5-generated two-step free nilpotent groups, the spectral sequence has been seen to collapse; that is, all the higher differentials vanish on all the generators of the cohomology of the exterior complex. Recent work [3] shows that there is an example of a 4-stage group which has cohomology differing from the cohomology of the associated exterior complex. To analyze such examples over Z requires more sophisticated machinery allowing iteration of the method above. Such methods are developed in [14].

REFERENCES

1. É. Cartan, Sur les invariants intégraux de certains espaces homogènes clos, Ann. Soc. Pol. Math., 8, (1929), 181–225.
2. H. Cartan and S. Eilenberg, Homological Algebra, Princeton University Press, Princeton, N.J. (1946).
3. B. Cenkl and R. Porter, Cohomology of nilmanifolds, Proc. Conf. on Algebraic Homotopy, Louvain, Lecture Notes in Math., Springer-Verlag, New York, in press.
4. C. Chevalley and S. Eilenberg, Cohomology theory of Lie groups and Lie algebras, Trans. Am. Math. Soc., 63, (1948), 85–124.
5. S. Eilenberg and J. Moore, Homology and fibrations I, Comment. Math. Helv., 40, (1966), 199–236.
6. L. Evens, Cohomology of Lie algebras in Pascal, Northwestern University, 1984.
7. V. K. A. M. Gugenheim and J. Peter May, On the theory and applications of differential torsion products, Mem. Am. Math. Soc., 142, (1974).

8. P. Hall, Nilpotent groups, Can. Math. Congress, Edmonton (1957) (reissued by Queens College, London).

9. A. Hearn, REDUCE User's Manual, Rand Corporation, Santa Monica, Calif. (April 1985).

10. J. Hübschmann, Homology and cohomology of nilpotent groups of class 2, preprint (1986).

11. J. L. Koszul, Homologie and cohomologie des algèbres de Lie, Bull. Soc. Math. France, 78, (1950), 65–127.

12. L. Lambe, Cohomology of principal G-bundles over a torus when $H^*(BG;R)$ is polynomial, Bull. Soc. Math. Belg., Volume in honor of Guy Hirsch, 38, (1986), 247–216.

13. L. Lambe and S. Priddy, Cohomology of nilmanifolds and torsion-free nilpotent groups, Trans. Am. Math. Soc., 273, (1982), 39–55.

14. L. Lambe and J. Stasheff, Applications of perturbation theory to iterated fibrations, Manuscripta Math., in press.

15. S. MacLane, Homology, Grundlehren Math. Wisc., Springer-Verlag, Berlin (1967).

16. E. Schruefer, EXCALC, a system for doing calculations in the calculus of modern differential geometry (September 1985).

17. R. Steinberg, Lectures on Chevalley groups, Yale University, Lecture Notes (1967).

11

Self-Similarity and Hairiness in the Mandelbrot Set

JOHN MILNOR

Institute for Advanced Study
Princeton, New Jersey

This chapter presents a conjectural description of the Feigenbaum limit of iterated period doubling, and its generalization to iterated period p-tupling, in terms of the geometry of the Mandelbrot set.

1. INTRODUCTION

Following Douady and Hubbard, the Mandelbrot set M is defined to be the set of all complex parameter values μ such that the orbit of $z = 0$ under the quadratic map

$$q_\mu : z \mapsto z^2 + \mu$$

is bounded (Figure 1). It is easy to show that this set M is compact, with connected complement. A theorem of [Douady, theorem 5] and Hubbard asserts that M is connected. Its detailed structure is extremely complicated. One very striking known property of the Mandelbrot set is the following: Any neighborhood of a boundary point of M contains infinitely many embedded copies of M. (Compare Figure 2. See Section 2 together with Appendix A for a precise statement and references.) Since M has interior points, it follows that the closure of the interior of M is equal to the entire set M.

By definition, a point μ in M is called a Misiurewicz point if the orbit of zero under the quadratic map q_μ is eventually periodic but not

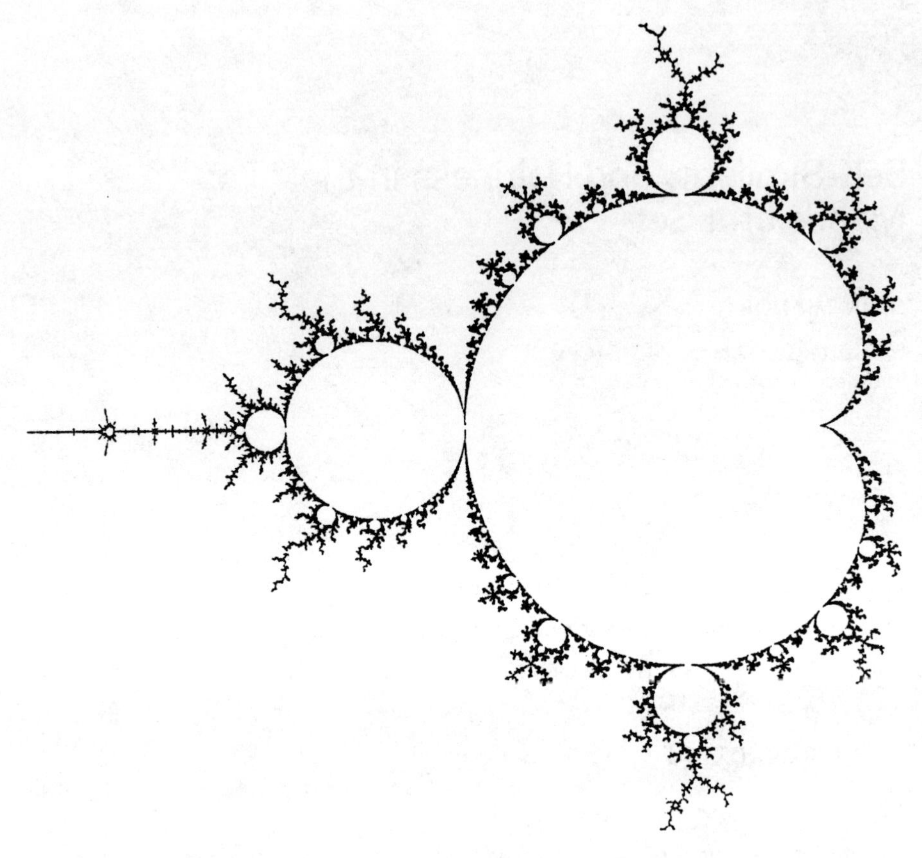

Figure 1 The boundary ∂M of the Mandelbrot set, slightly thickened so that very fine hairs are visible.

periodic. Here is another striking property. The Mandelbrot set is "self-similar" about any Misiurewicz point μ in a sense which can be described roughly as follows: If we examine a neighborhood of μ in M with a very powerful microscope, and then increase the magnification by a carefully chosen factor, the picture will be unchanged except for a rotation (Figure 3; see Appendix A for a precise statement). There are many Misiurewicz points. In fact, they form a countable dense subset of the boundary of M.

This chapter is concerned with possible self-similarity at a different set of boundary points of M. First a preliminary definition. It will be convenient to say that a point s of M is superstable of period $p \geq 1$ if the orbit of zero under q_s is strictly periodic with period equal to p. There are finitely many such points for each positive integer p, and

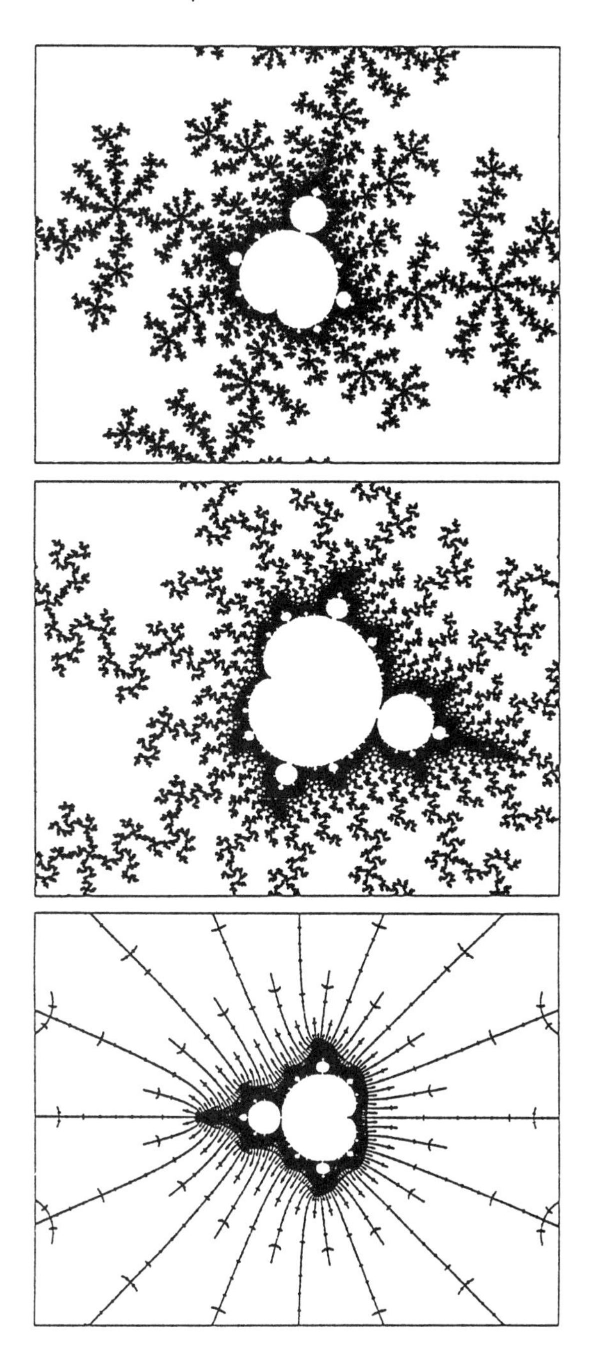

Figure 2 Small copies of M in M. These particular pictures are centered at 3 of the 1013 distinct primitive hyperbolic components of period 11 in the Mandelbrot set. Again only the boundary of M is shown. (The pictures are centered at (−0.669,0.475), (0.019,.0.763), and −1.9308, and the widths of the boundary frames are 0.0021, 0.000085, and 0.0000013, respectively.)

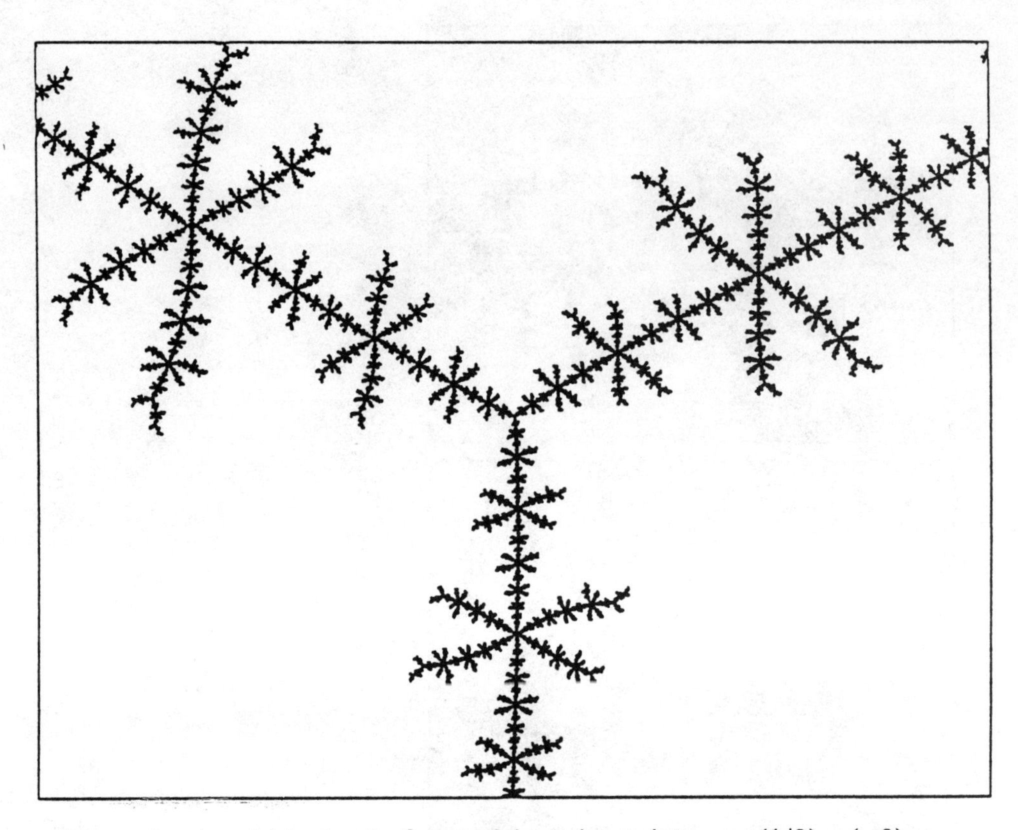

Figure 3 A neighborhood of the Misiurewicz point $\mu = s(1/3) \cdot (-2) = -0.1011 + 0.9563i$, which is conspicuous near the top of Figure 1, magnified some 10,000 times. If we further magnify by a complex factor of $\lambda = -0.655 + 1.156i$, or in other words if we magnify by $|\lambda| = 1.328$ and then rotate by $\arg(\lambda) = 119.6^0$ to the left, then this picture will remain unchanged. Here μ is chosen so that $q_\mu^{04}(0)$ is a fixed point of q_μ, and λ is equal to the derivative of q_μ at this fixed point. (Compare Appendix A. Frame width: 0.00030.)

their number grows exponentially with p. Given a superstable point s of period $p > 1$, one can construct an associated infinite sequence of distinct superstable points

$$0,\ s,\ s*s,\ s*s*s,\ \cdots,$$

where the n-fold star product, which we will write as s^{*n}, has period equal to p^n. (See Section 2.) All of the known examples are compatible with the following statement.

Conjecture 1.1

This sequence of points s^{*n} always converges geometrically.

In other words the sequence of difference ratios $(s^{*n} - s^{*n-1})/(s^{*n+1} - s^{*n})$ converges as $n \to \infty$ to some complex constant δ with $|\delta| > 1$. Hence the sequence $\{s^{*n}\}$ itself converges to a limit, denoted briefly by s^{∞}. We will call this limit s^{∞} a generalized Feigenbaum point. Here are three examples.

p	s	s^{*2}	s^{*3}	s^{*4}	...	s^{∞}	δ
2	-1	-1.310703	-1.381547	-1.396945	...	-1.401155189	4.669201609
3	-.122561 +.744862 i	-.031553 +.790783 i	-.023369 +.784680 i	-.023539 +.783683 i	...	-.023641169 +.783660651 i	4.6002246 +8.9812259 i
3	-1.754878	-1.785866	-1.786430	-1.786440	...	-1.786440256	55.2470266

In the first case $s = -1$, Feigenbaum observed empirically that the sequence of points s^{*n} of period 2^n converges geometrically; and this statement has been proved by Collet et al., Campanino and Epstein, and Lanford. (Compare the papers collected in Cvitanovic [2].) The next case, with period 3, is one of an infinite family indexed by roots of unity, studied by Golberg et al. [13] and by Cvitanovic and Myrheim. The third case is one of an infinite sequence converging towards -2, studied by Eckmann et al. [10]. Further examples may be found in Appendix C.

Given such a generalized Feigenbaum point s^{∞}, the problem studied here is the following. Is M self-similar about s^{∞}, with magnification ratio equal to δ? That is, if we look at a highly magnified picture of M in a neighborhood of s^{∞}, and then increase the magnification by a factor of δ, will the picture be essentially unchanged? (See Figures 4 and 5.) In fact, the answer seems to depend on just how these pictures are drawn. Computer experiments suggest the following three statements. Let $(M - s^{\infty})\delta^n$ stand for the set of all products $(\mu - s^{\infty})\delta^n$ with $\mu \in M$.

Conjecture 1.2

The successive magnified images $(M - s^{\infty})\delta^n$ converge in measure to some limit set X, within any bounded region of the plane. That is,

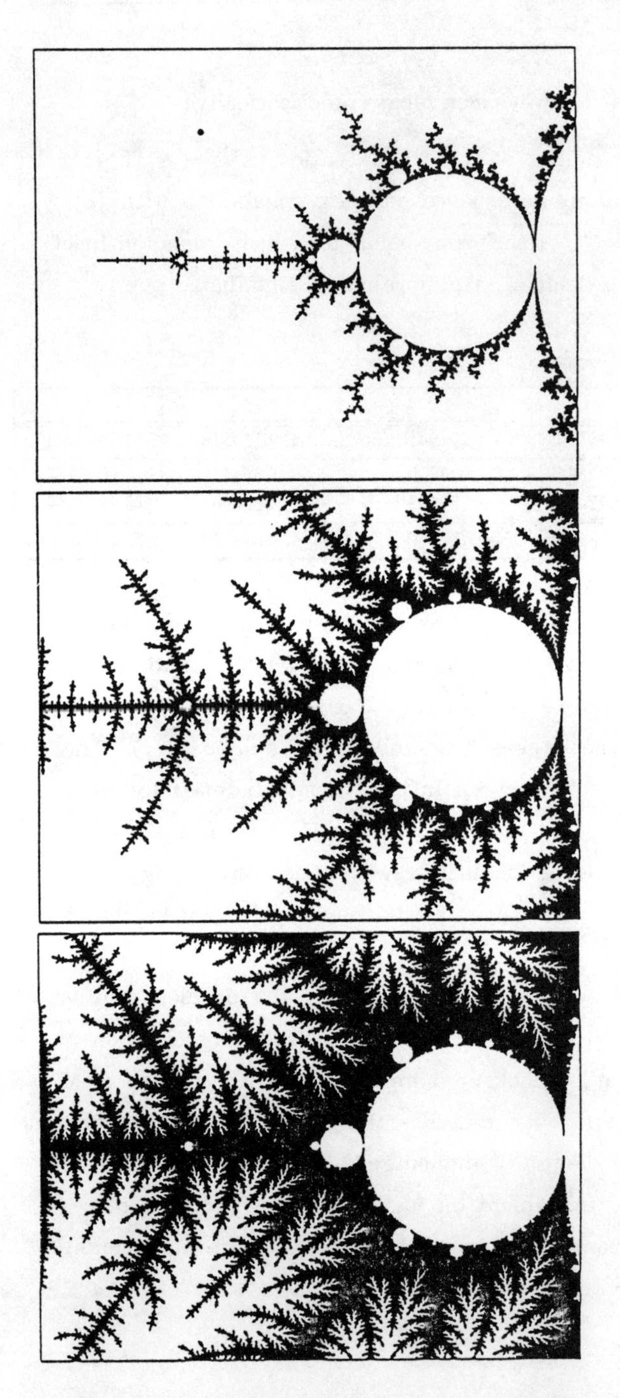

Figure 4 Three successive pictures of ∂M centered at the classical Feigenbaum point $(-1)^\infty = -1.401155$. The first of these three shows the entire left half of M. Each succeeding picture has been magnified by a factor of $\delta^2 = 21.801$. (Frame widths: 1.55, 0.071, 0.0033.)

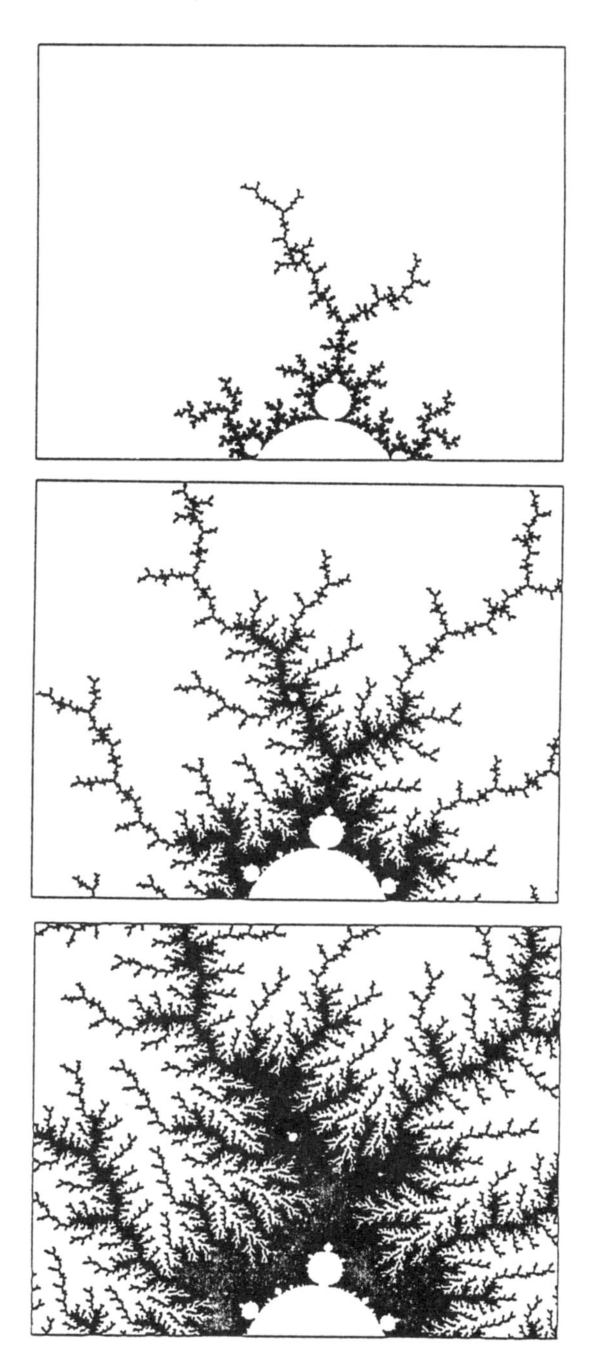

Figure 5 Three successive pictures centered at the generalized Feigenbaum point $s^{\infty} = -0.1528 + 1.0397i$, where $s = -0.1565 + 1.0322i$ is the uppermost period 4 superstable point. The first of these pictures extends from the top of M down to the top portion of the period 3 component $s(1/3) \cdot C_0$. Each succeeding picture has been magnified by a factor of $\delta = 100.4 + 69.3i$. (Frame widths: 0.64, 0.0052, 0.000043.)

there exists a measurable set $X \subset C$ so that the Lebesgue measure of the symmetric difference of $(M - s^\infty)\delta^n$ and this limit set X, intersected with any fixed bounded region of the complex plane, tends to zero as $n \to \infty$.

A closely related and more explicit conjecture will be given in Conjecture 3.3.

Conjecture 1.3

The limit set X is not a closed set, but rather is everywhere dense in the complex numbers C.

Without assuming Conjecture 1.2, we could simply conjecture that the successive magnified images $(M - s^\infty)\delta^n$ fill out the complex plane more and more densely, so that an arbitrary disk, anywhere in C, must intersect $(M - s^\infty)\delta^n$ for every n which is sufficiently large. (Compare Section 6.) If $r_\varepsilon(\mu)$ is defined to be the radius of the largest open disk which is disjoint from M and lies inside the ε-neighborhood of μ, then an equivalent statement would be that the ratio $r_\varepsilon(s^\infty)/\varepsilon$ tends to zero as $\varepsilon \to 0$. In this form, one could make the same conjecture at many other boundary points of M. (Compare Figure 8.)

Conjecture 1.4

This limit set X is extremely sparse in some regions, so that for any $\varepsilon > 0$ there exists a unit disk in C whose intersection with X has measure less than ε.

In fact, it seems empirically that the successive magnified images $(M - s^\infty)\delta^n$ fill out the plane by growing many fine hairs, most of which have extremely small measure.

If we want to draw a picture of the Mandelbrot set M with a computer, the naive procedure might be to color a pixel, say, black if its midpoint belongs to M and white otherwise. However, such a picture would be very misleading since it would completely ignore the many filaments of small measure. If three colors are available, then a much better procedure is to choose color 1, 2, or 3 according as the pixel lies completely in the interior of M, completely outside M, or contains a point of the boundary ∂M. If we adopt this convention, then Conjecture 1.3 asserts that a highly magnified picture near a generalized Feigenbaum point must contain only pixels with colors 1 and 3. No pixel, in such a highly magnified picture, can lie completely outside M.

Since only two colors are available for this publication, I have adopted a third convention, attempting to color a pixel black if it contains a boundary point of M and white in the other two cases. I do not know any efficient algorithm for deciding definitely whether a pixel contains a boundary point or not. However, the algorithm which was used, based on Lemma 5.6 below, is expected to provide a good approximation to the correct picture. The computations were carried out, and the figures produced, on a Sun Workstation.

Related conjectures will be discussed in Sections 3, 4, and 7. Although much is known about the geometry of M, particularly due to the work of Douady, Hubbard, and Thurston, there are still many open questions. The most fundamental unsolved problem is the following. [Compare (5.3).]

Conjecture 1.5

M is locally connected.

Douady and Hubbard [7] have shown that 1.5 implies the following.

Conjecture 1.6

For every interior point μ of M, the orbit of zero under the quadratic map q_μ must converge toward a stable periodic orbit.

An equivalent statement would be that every point μ_0 of M can be approximated arbitrarily closely by points μ for which q_μ has a stable periodic orbit (Section 2). Note that this is unknown even in the real case. It is surely true, although unproved, that every point μ_0 in the interval $M \cap R = [-2, 1/4]$ can be approximated arbitrarily closely by real points μ for which q_μ has a (necessarily real) stable periodic orbit.

Another interesting question is the following. (Compare Mandelbrot [17].)

Conjecture 1.7

The boundary of the Mandelbrot set has Hausdorff dimension equal to 2.

2. THE SEMIGROUP OF SUPERSTABLE POINTS

Let $f : C \to C$ be a complex analytic map. The nth iterate $f \circ \cdots \circ f$ will be denoted by $f^{\circ n}$. A periodic orbit, $f^{\circ p}(z) = z$, is said to be either stable, unstable, or indifferent according as the "characteristic derivative" $\lambda = (f^{\circ p})'(z)$ around this orbit has absolute value less than, greater than, or equal to one. In the case of a quadratic map, a classical theorem of Fatou and Julia asserts that there can be at most one stable periodic orbit. In fact, the successive images of the unique critical point must necessarily converge to a stable orbit, whenever one exists.

A periodic orbit is called superstable if it actually contains a critical point of the function, so that the characteristic derivative is precisely zero. As noted in Section 1, we extend this terminology by calling a parameter value s in the Mandelbrot set "superstable" whenever the corresponding quadratic map q_s possesses a superstable orbit, that is, whenever its unique critical point $z = 0$ is contained in a periodic orbit. Each of the small copies of M in M, which were mentioned in Section 1, can be uniquely labeled by its superstable point of smallest period. In this way, we obtain a semigroup of embeddings of M into itself, labeled by the collection of superstable points. More precisely, we have the following assertions, due to Douady and Hubbard (who use the notation $s \perp \mu$ in place of our $s * \mu$). Parts of these statements may be found in Douady [5] and Douady and Hubbard [8], but some are not yet published.

Assertion 2.1

To each superstable point $s \in M$ there is associated a topological embedding, which I will denote by $\mu \mapsto s * \mu$, from the Mandelbrot set M into itself.

Thus the image $s * M \subset M$ is a homeomorphic copy of the full Mandelbrot set. On a rough intuitive level, this Douady-Hubbard product $s * \mu$ can be characterized by the following two properties.

The dynamic behavior of the quadratic map $q_{s*\mu}$ is qualitatively similar to that of q_s, so long as we stay away from a suitable neighborhood of the orbit of zero; but:

Within this neighborhood, the p-fold iterate of $q_{s*\mu}$, where p is the period of s, behaves dynamically like q_μ.

A more precise characterization will be given following 2.9 below. (See also 5.5 and Appendix B.) Douady and Hubbard refer to $s * \mu$ as s "modulated" or "tuned" by μ and describe $s * M$ as a small and somewhat distorted copy of the Mandelbrot set which is "subordinated" to the superstable point s. Here are some basic properties.

Property 2.2

The unique superstable point 0 of period one serves as two-sided identity element, so that $0 * \mu = \mu$ and $s * 0 = s$ for every μ and s.

Property 2.3

The point $s * \mu$ is superstable if and only if μ is superstable. More generally, the map $q_{s*\mu}$ has a stable periodic orbit if and only if q_μ has a stable periodic orbit. Furthermore, the period of the stable orbit for $q_{s*\mu}$ is equal to the product of the corresponding periods for q_s and q_μ, and the characteristic derivative around the stable orbit for $q_{s*\mu}$ is equal to the corresponding characteristic derivative for q_μ.

Thus $0 * M$ is equal to the full Mandelbrot set, but the embedded copy $s * M$ is a proper subset of M for $s \neq 0$. In fact, most of the sets $s * M$ are extremely small. For any $\varepsilon > 0$ it is surely true, although not proved, that all but finitely many of these sets $s * M$ have diameter less than ε.

Property 2.4

The product $s * \mu$ is a Misiurewicz point if and only if μ is a Misiurewicz point.

For the relationship between the eventual periods of the Misiurewicz points μ and $s * \mu$, see Appendices A and B. Since the closure of the set of Misiurewicz points is the boundary of M, it follows that

The product $s * \mu$ is either an interior or a boundary point of M according as μ is an interior or a boundary point of M.

Using this product operation, we have the following statement.

Property 2.5

The countably infinite set S consisting of all superstable points in the Mandelbrot set is a free monoid, acting on the space M. That is, S is a noncommutative semigroup with identity element; the associative law $(s * s') * \mu = s * (s' * \mu)$ is always satisfied; and any nonidentity element of S can be factored uniquely as a product of indecomposable elements. Furthermore, if s and s′ are distinct and indecomposable, then the compact sets $s * M$ and $s' * M$ are disjoint from each other.

In order to state a further property, we will need the following [8].

Definition 2.6

A complex analytic function f, defined on a bounded simply connected open set $N \subset C$, is polynomial-like of degree d if f maps N onto an open neighborhood f(N) of its closure $\overline{N}$ by a proper map of degree d. In the special case $d = 2$, we will say briefly that f is quadratic-like.

Thus f maps points near the boundary of N to points which lie near the boundary of f(N) and hence are outside $\overline{N}$. Every point of f(N) has exactly d preimages in N if these are counted with the appropriate (always positive) multiplicity. It follows also that N contains exactly $d - 1$ critical points of f, counted with multiplicity.

Definition 2.7

The "filled-in Julia set" $K(f) \subset N$ is the compact set consisting of all points $z \in N$ whose orbits under f stay completely within N, so that $f^{on}(z) \in N$ is defined for all $n \geq 0$.

(The boundary $\partial K(f)$ is called the Julia set.) Douady and Hubbard show that this set $K(f) = K(f \mid N)$ is connected if and only if it contains all of the critical points of f in N. Note that any polynomial F of degree d is automatically polynomial-like of degree d in a suitably chosen large region. We can simply take N to be the inverse image under F of a large disk centered at the origin. In this case, the set $K(F \mid N)$ does not depend on the particular choice of N, as long as N is large enough, and we will use the abbreviated notation $K(F)$.

Definition 2.8

Two polynomial-like maps f and g are quasi-conformally equivalent if there exists a quasi-conformal homeomorphism ϕ from a neighborhood of $K(f)$ onto a neighborhood of $K(g)$ so that $g = \phi \circ f \circ \phi^{-1}$. Douady and Hubbard show that:

> Any quadratic-like map is quasi-conformally equivalent to some polynomial of degree 2, and hence to some q_μ. Furthermore, if μ is a superstable point or if μ lies in the boundary of the Mandelbrot set, then it is uniquely determined.

That is, in this case, q_μ cannot be quasi-conformally equivalent to any q_ν with $\nu \neq \mu$. A basic property of the Douady-Hubbard product $\eta = s * \mu$ can now be described as follows.

Property 2.9

Suppose that the p-fold iterate $f = q_\eta^{\circ p}$ of the quadratic map q_η is quadratic-like in some neighborhood N of the origin. Then η can be expressed as a product $s * \mu$ with s of period p and with q_μ quasi-conformally equivalent to $f \mid N$. Conversely, if η can be expressed as $s * \mu$ with s of period p and with $\mu \neq 1/4$, then $f = q_\eta^{\circ p}$ is quadratic-like in some simply connected domain N which contains the origin and which varies continuously with $s * \mu$, and this quadratic-like map $f \mid N$ is quasi-conformally equivalent to q_μ.

Here we must exclude the point $\mu = 1/4$, also known as the "root point" of M, since it sometimes is badly behaved. If $N = N_0 \to N_1 \to N_2 \to \cdots$ are the successive images of N under $q_{s*\mu}$, then evidently the set $N = N_0$ must be symmetric about the origin, in the sense that $N = -N$, and must contain the entire orbit of zero under f. On the other hand, each of the sets N_k with $0 < k < p$ must be disjoint from $-N_k$, so that N_k maps diffeomorphically onto N_{k+1}. Finally, the set

N_p must contain the closure of N_0. (Compare Figure 6.) In nearly all of the cases I have checked, one can choose the neighborhood $N = N_0$ to be the interior of a carefully chosen ellipse. It is often not possible to choose N so that the sets $N_0, \ldots, N_{p-1}$ are disjoint from each other.

If we are willing to assume Conjecture 1.6, then the various properties which we have described do provide a precise (but nonconstructive) characterization of the Douady-Hubbard product $s * \mu$. Recall from 2.5 together with continuity that the various image sets $s * M$, with s indecomposable, are disjoint, compact, and connected. Assertion 2.9 tells us how to test whether some given point of M belongs to the union of the sets $s * (M - 1/4)$ with $s \neq 0$. Within each connected component of this union, note that the indecomposable first factor s can be characterized simply as the superstable point of smallest period. The second factor $\mu \neq 1/4$ is uniquely determined by 2.9 when it is a boundary point or a superstable point, and by 2.3 together with continuity in the other cases. Finally, the case of a product of the form $s * 1/4$ is easily handled by continuity.

Missing in the above discussion is any effective recipe for computing the product $s * \mu$. In order to give a more explicit description, it is almost necessary to have some more conceptual labeling of points in M, and particularly of superstable points. One method of doing this, based on the Hubbard tree, is described in Appendix B. Another method, based on the concept of external argument, is described in Section 5. Still another method, based on the concept of geodesic lamination, may be found in Thurston [20] and Levy [15]. A few further comments about the construction of the Douady-Hubbard product will be given at the end of this section.

Using this product operation, we can begin to describe the structure of the Mandelbrot set as follows. Let M_0 be the open set consisting of all complex parameters μ such that the quadratic map q_μ has some stable periodic orbit. Douady and Hubbard show that this open set $M_0 \subset M$ splits up into countably many connected components, each of which is canonically biholomorphic to the open unit disk D, using the characteristic derivative $\lambda = \lambda(\mu)$ around the stable orbit for q_μ as canonical uniformizing parameter. They call these sets hyperbolic compo-

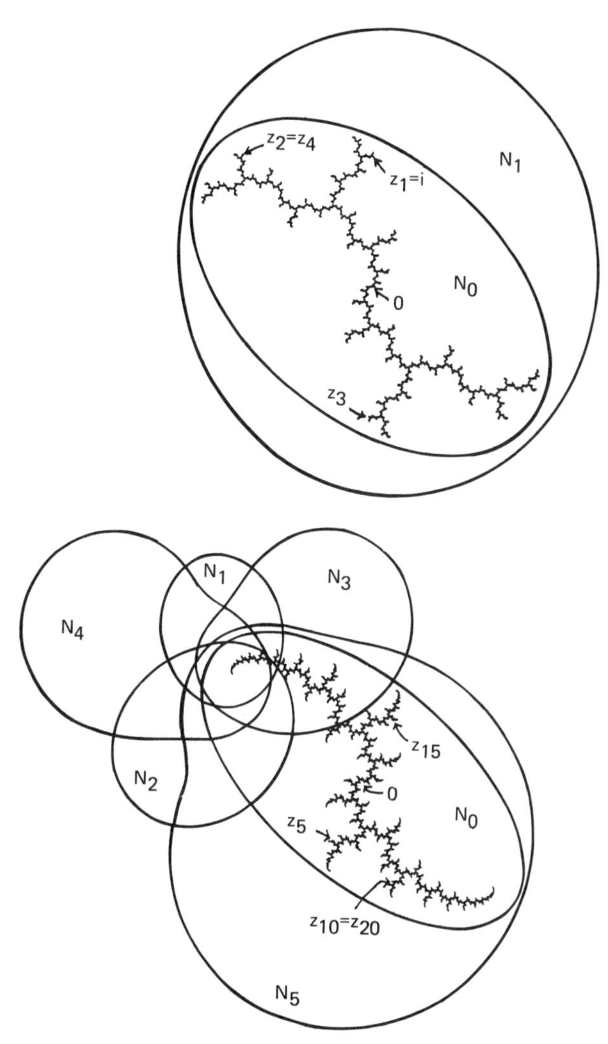

Figure 6 At the top is a picture of the Julia set $K(q_i) = \partial K(q_i)$ associated with the Misiurewicz point $i = \sqrt{-1}$. Below is the Julia set for the 5-fold iterate of the quadratic map q_μ, where $\mu = s(2/5) \cdot i$, restricted to a carefully chosen neighborhood N_0. The successive images of the appropriate neighborhood N_0 are shown in both cases, and the orbit $0 \to z_1 \to z_2 \to \cdots$ of the critical point is shown in so far as it lies in N_0.

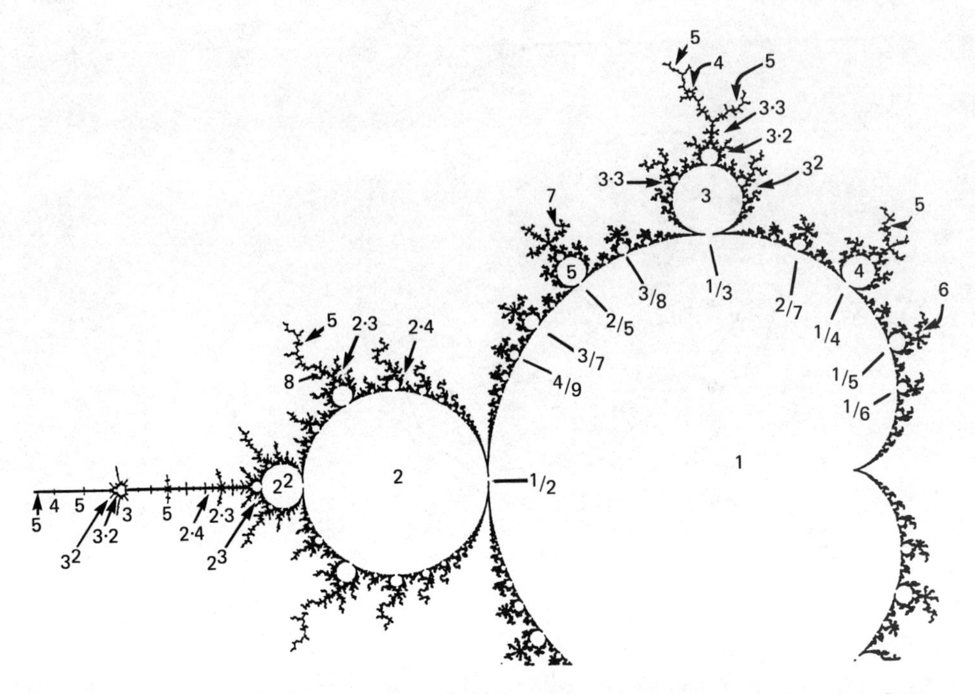

Figure 7 Annotated Mandelbrot set, showing the periods of some of the more prominent hyperbolic components in the upper half of M. All those of period five or less are indicated, even though some are not really visible at this scale. The ratios m/p for some of the satellites of C_0 are also shown.

nents of the interior of M. (See Figures 7 and 8, as well as Appendix C.) In particular, it follows that each hyperbolic component contains a unique superstable point s, which they call its "center." The period of the stable orbit is constant throughout each hyperbolic component.

We will use the notation C_s for the hyperbolic component which contains the superstable point s. Thus there is a canonical diffeomorphism $\lambda : C_s \to D$, with $\lambda(s) = 0$. As noted in 1.6, it is conjectured that the union M_0 of the hyperbolic components C_s is equal to the entire interior of M. It follows from 2.3 that the map $\mu \mapsto s * \mu$ carries each hyperbolic component C_t biholomorphically onto the hyperbolic component $s * C_t = C_{s*t}$.

The hyperbolic component C_0 containing the origin, s = 0, is particularly conspicuous and easy to describe. It consists of all parameter values μ such that q_μ has a stable fixed point. Since the derivative at

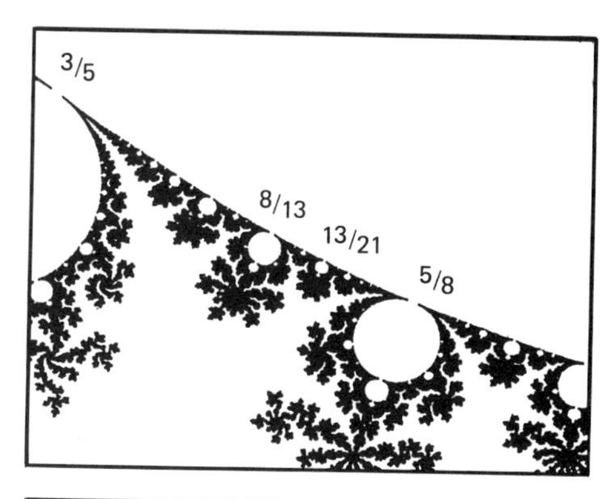

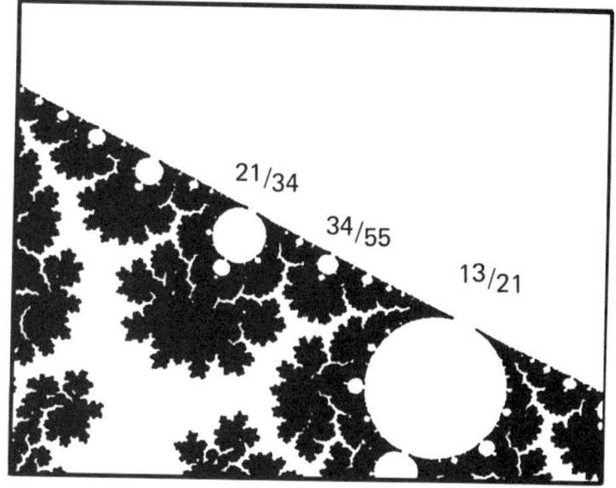

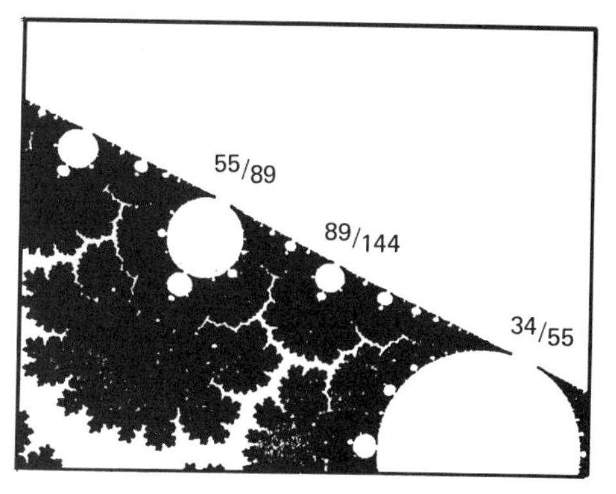

Figure 8 Three successive pictures centered at the point $\mu(e^{2xi\alpha})$ on the boundary of the principal hyperbolic component C_0, where $\alpha = (\sqrt{5} - 1)/2$ cannot be approximated closely by rationals. Mandelbrot [17] has used the increasing density of boundary points in such figures as support for the conjecture that ∂M has Hausdorff dimension two. (Frame widths: 0.2, 0.02, 0.002. The approximating rationals are ratios of Fibonacci numbers.)

a fixed point $z = z^2 + \mu$ is given by $\lambda = q_\mu'(z) = 2z$, we can solve for μ as a function $\mu = \hat{\mu}(\lambda)$, or conversely for $\lambda = \hat{\lambda}(\mu)$, where the function $\hat{\mu}$ and its inverse function $\hat{\lambda}$ are given by

$$\hat{\mu}(\lambda) = \lambda(2 - \lambda)/4, \qquad \hat{\lambda}(\mu) = 1 - \sqrt{1 - 4\mu} \qquad [2.10]$$

as λ ranges over the open unit disk D and μ varies over $C_0 = \hat{\mu}(D)$. Thus the principal hyperbolic component C_0 is bounded by the cardioid $\partial C_0 = \hat{\mu}(\partial D)$ consisting of all points $e^{i\theta}(2 - e^{i\theta})/4$. This cardioid is a smooth curve, except for a cusp at the "root point" $\hat{\mu}(1) = 1/4$. This root point can be characterized as the unique parameter value μ such that the quadratic map q_μ has just one fixed point, with characteristic derivative $\lambda = 1$.

It is sometimes useful to apply the map $\hat{\lambda}$, defined by (2.10), to the entire Mandelbrot set (Figure 9). We will think of $\hat{\lambda}$ as a conformal diffeomorphism from the plane of complex numbers μ, slit along the line $[1/4,\infty)$, to the half-plane $\mathrm{Re}(\lambda) < 1$. Note that the two numbers $\hat{\lambda}(\mu)$ and $2 - \hat{\lambda}(\mu)$ can be described as the characteristic derivatives of the quadratic map q_μ at its two fixed points.

For $s = -1$ one has the explicit formula $(-1) * \hat{\mu}(\lambda) = -1 + \lambda/4$ for $|\lambda| \leq 1$, so that the hyperbolic component $C_{-1} = (-1) * C_0$ is a round disk. There is no such easy formula in general.

Particularly conspicuous in any picture of the Mandelbrot set are those hyperbolic components $C_s = s * C_0$ of period $p > 1$ which are immediately contiguous to the principal component C_0 in the sense that the boundary curve ∂C_s is tangent to the cardioid ∂C_0. The point of tangency is necessarily equal to the root point $s * \hat{\mu}(1)$ of C_s and must be of the form $\hat{\mu}(\omega)$, where ω is a primitive pth root of unity. If ω is equal to $e^{2\pi i m/p}$, with $0 < m/p < 1$, then we will use the notation $s = s(m/p)$ for the corresponding contiguous superstable point, with root point

$$s(m/p) * \hat{\mu}(1) = \hat{\mu}(e^{2\pi i m/p}) \qquad [2.11]$$

The associated hyperbolic component $C_{s(m/p)}$ has smooth boundary, even at this root point, and can be closely approximated by a round disk of radius $\sin(\pi m/p)/p^2$. The picture becomes somewhat easier to understand if we apply the change of coordinates (2.10), as in Figure 9.

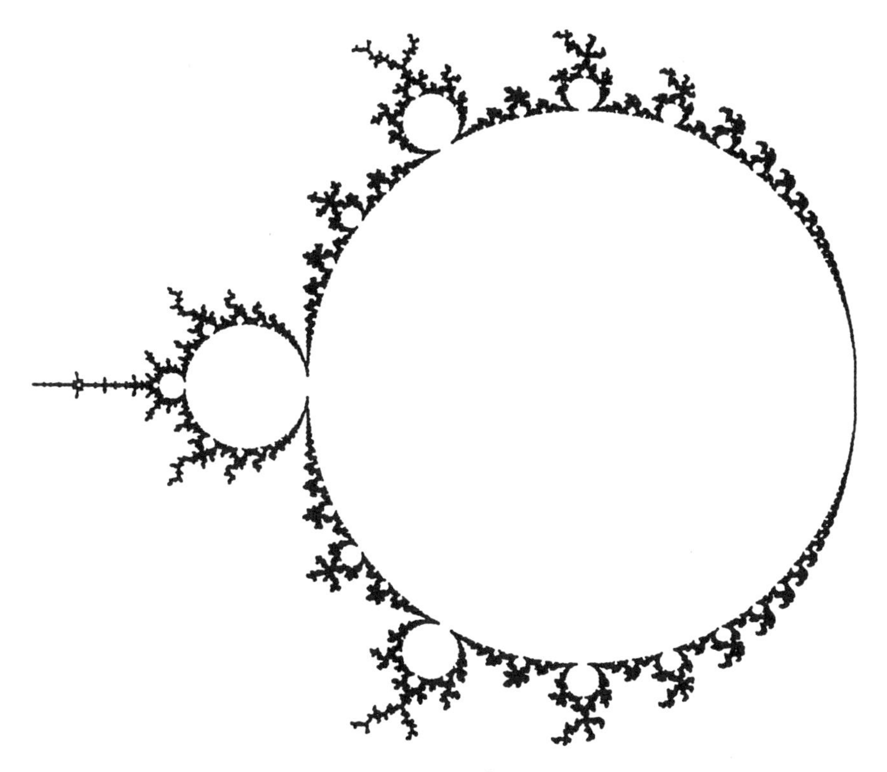

Figure 9 Image of ∂M under the map $\hat{\lambda}$ of Section 2.10, which maps the principal component C_0 onto the unit disk, and thus straightens out its cusp point. If $\lambda = \hat{\lambda}(\mu)$, then the map $q_\mu : z \to z^2 + \mu$ is equivalent under a linear change of variable to the map $\zeta \to \lambda\zeta + \zeta^2$, or (when $\lambda \neq 0$) to the "logistic map" $w \to \lambda w(1 - w)$.

Then the image $\hat{\lambda}(C_{s(m/p)})$ is approximately a round disk of radius $1/p^2$ and is attached to the unit disk $D = \hat{\lambda}(C_0)$ precisely at the root of unity $e^{2\pi i m/p}$. (Compare [Mandelbrot] as well as [Cvitanovic and Myrheim].) The period p of this contiguous hyperbolic component can be read off easily from either picture, since there is a conspicuous p-branched spiral appended to $C_{s(m/p)}$, with center at the Misiurewicz point $s(m/p) * (-2)$.

Douady and Hubbard call $C_{s(m/p)}$ the satellite of C_0 with argument m/p. More generally, they divide the hyperbolic components C_s, or equivalently the superstable points s, into two classes, depending on behavior at the root point $s * (1/4)$. The component C_s is said to be primitive if it satisfies the following equivalent conditions:

1. The boundary curve $\partial C_s = s * \hat{\mu}(\partial D)$ has a cusp at its root point, so that the embedded copy $s * M$ of the Mandelbrot set looks qualitatively very much like the full Mandelbrot set.
2. The quadratic map $q_{s*(1/4)}$ associated with the root point has a periodic orbit with characteristic derivative $\lambda = 1$ and with period equal to the period p of s.
3. The superstable point s cannot be factored as a product of the form $s' * s(m/r)$.

Alternatively, C_s is a satellite of some other component $C_{s'}$ if:

1'. Its boundary ∂C_s is a smooth curve, even at the root point, so that the embedded copy $s * M$ of the Mandelbrot set looks qualitatively more like Figure 9;

2'. The quadratic map $q_{s*(1/4)}$ has a periodic orbit with characteristic derivative λ equal to some nontrivial root of unity $e^{2\pi i m/r}$ and with period equal to some proper divisor p/r of the period p of s; or equivalently

3'. The superstable point s can be factored as $s' * s(m/r)$, hence the boundary curves of the two hyperbolic components C_s and $C_{s'}$ are tangent to each other at the root point $s * (1/4) = s' * \hat{\mu}(e^{2\pi i m/r})$.

The number of superstable points s (or of hyperbolic components C_s) of period p can be computed as follows. First note that there are exactly 2^{n-1} distinct s for which the period p is a divisor of n. These s are just the roots of the polynomial $s \mapsto q_s^{\circ n}(0)$, which has degree 2^{n-1}. This polynomial has nonnegative integer coefficients, with leading coefficient one. Thus every s is an algebraic integer, and in fact every $s \neq 0$ is an algebraic unit. These roots are all distinct, since it is easy to check that the first derivative of the polynomial, evaluated at any s, is congruent to one modulo two and hence is nonzero. It follows that the number of distinct superstable points of period p is equal to

$$\sum 2^{p/d} \mu(d)/2 \qquad [2.12]$$

to be summed over all square-free divisors $d|p$. Here $\mu(d)$ stands for the Moebius function, equal to $(-1)^m$ when d is a product of m distinct primes. This number (2.12) is asymptotic to 2^{p-1} as $p \to \infty$. From (2.12), it is not difficult to compute the precise number of monoid elements of period p which are primitive, or indecomposable, or both. These numbers are also asymptotic to 2^{p-1} as $p \to \infty$. As examples, for p equal to 2 through 16, the number of indecomposables is equal to

1, 3, 5, 15, 21, 63, 107, 243, 465, 1023, 1969, 4095, 8007, 16287, 32248

respectively. The number of real superstable points can be computed as $\sum 2^{p/d}\mu(d)/2p$, to be summed over odd square-free divisors of p. (Compare [18].) This number is asymptotic to $2^{p-1}/p$.

We conclude this section with some remarks on the computation of $s * \mu$. If we restrict attention to the case where μ is either superstable or a Misiurewicz point, then the complex number $s * \mu$ satisfies a polynomial equation with integer coefficients. Hence it is always possible to use Newton's method to find the solution to very high precision, once we know a good approximation to it. If s is primitive, and if the degree of μ is not too high, then the linear Taylor series about $\mu = 0$ seems to give a sufficiently good approximation. This has the form

$$s * \mu = s + \mu/a(s) + (\text{higher terms}) \qquad [2.13]$$

where the coefficient a(s) is equal to the product of the first derivative of the correspondence $\mu \mapsto q_\mu^{\circ p}(0)$ at $\mu = s$ and the first derivative of $z \mapsto q_s^{\circ p-1}(z)$ at $z = s$. However, if s is a satellite, then this correspondence $\mu \mapsto s * \mu$ has a bad singularity at $\mu = 1/4$. This singularity can be eliminated by using the change of coordinates (2.10). In the case $s = s(m/p)$, the linear Taylor approximation about the root point $\hat{\lambda}(\mu) = 1$ then takes the explicit form

$$\hat{\lambda}(s(m/p) * \mu) = e^{2\pi i m/p} + (1 - \hat{\lambda}(\mu))\, e^{2\pi i m/p}/p^2 + (\text{higher terms}) \qquad [2.14]$$

(Compare [Cvitanovic and Myrheim]. In the special case $s = -1$ the approximation $(-1) * \mu \approx -1 + \hat{\lambda}(\mu)/4$ is even better.) Another useful approximation will be given in (3.2).

3. GENERALIZED FEIGENBAUM POINTS

Let s be an arbitrary nonzero element of the monoid $S \subset M$ of superstable points, and let s^{*n} denote the n-fold product $s * \cdots * s$. According to Conjecture 1.1, this sequence of points s^{*n} must converge to a limit, which we denote by s^∞ and refer to as a generalized Feigenbaum point. (More generally, for any infinite sequence $s_1, s_2, \ldots$ of nonzero monoid elements it seems natural to conjecture that the finite products $s_1 * \cdots * s_n$ converge as $n \to \infty$ to a unique boundary point of the Mandelbrot set, which can be described as an infinite product.) We now sharpen Conjecture 1.1 as follows.

Conjecture 3.1

The infinite product s^∞ is the unique fixed point of the correspondence $\mu \mapsto s * \mu$. Furthermore, this correspondence has a well-defined derivative, with absolute value less than one, at the fixed point s^∞.

Evidently this derivative must be just the reciprocal of the self-similarity constant δ of Conjecture 1.1. Hence the Taylor expansion about s^∞ takes the form

$$s * \mu \approx s^\infty + (\mu - s^\infty)/\delta \qquad [3.2]$$

Caution. The correspondence $\mu \mapsto s * \mu$ is definitely not differentiable in general. For example, it is quite likely never differentiable at a Misiurewicz point. Differentiability is conjectured only at its fixed point s^∞.

An even sharper version can be stated as follows. Fixing the superstable point $s \neq 0$, for each positive integer n consider the embedding $\mu \mapsto (s^{*n} * \mu - s^\infty)\delta^n$ from the Mandelbrot set into C.

Conjecture 3.3

As $n \to \infty$ these embeddings $\psi_n(\mu) = (s^{*n} * \mu - s^\infty)\delta^n$ converge uniformly to a topological embedding $\psi : M \to C$.

Thus it is conjectured that $s^{*n} * \mu = s^\infty + \psi(\mu)/\delta^n + o(1/\delta^n)$ as $n \to \infty$. Note that this limit embedding must satisfy the identity $\psi(s * \mu) =$

$\psi(\mu)/\delta$ for every μ. It follows that $\psi(s^{*n} * M) = \psi(M)/\delta^n$. Hence there are inclusions.

$$\cdots \subset \psi(M)/\delta^2 \subset \psi(M)/\delta \subset \psi(M) \subset \psi(M)\delta \subset \psi(M)\delta^2 \subset \cdots$$

The union of these compact sets $\psi(M)\delta^n$ should be identified with the limit set X of Conjecture 1.2, and is conjectured to be everywhere dense in the complex plane.

Unfortunately, I do not know any efficient algorithm for drawing a picture of the image set $\psi(M)$. Presumably such a picture would look like a slightly distorted version of Figure 1 in the case of a primitive superstable point and like a (somewhat more) distorted version of Figure 9 for a satellite. As a rough measure of distortion we can consider the difference ratio $(\psi(0) - \psi(-1))/(\psi(-1) - \psi(-2))$, which is relatively easy to compute. We would expect this ratio to be close to 1 if the superstable point s is primitive, but closer to $(\hat{\lambda}(0) - \hat{\lambda}(-1))/(\hat{\lambda}(-1) - \hat{\lambda}(-2)) \approx 1.62$ for a satellite. Here is a table which lists values for two satellites and for one primitive case.

p	s		$\psi(0)$	$\psi(-1)$	$\psi(-2)$	$(\psi(0)-\psi(-1))/(\psi(-1)-\psi(-2))$
2	-1	$= s(1/2)$	2.0023287	.42883749	-.63536011	1.4785705
3	-.12256 +.74486 i	$= s(1/3)$	-.118859 -1.076994 i	-1.1037 -.4331 i	-1.8665 +.1041 i	1.396 -.242 i
3	-1.75488		1.753337	.75195	-.21455	1.03610

4. THE CVITANOVIC-FEIGENBAUM OPERATOR

To every integer $p \geq 2$, we associate the functional operator T_p which is defined by

$$T_p : f(z) \mapsto g(z) = f^{\circ p}(\alpha z)/\alpha, \qquad \alpha = f^{\circ p}(0) \qquad [4.1]$$

whenever $f^{\circ p}(0) \neq 0$. Here f and $T_p(f) = g$ are functions of one complex variable with a critical point at the origin, and $f^{\circ p}$ stands for the p-fold iterate of f. This operator is normalized so that $g(0) = 1$. (The proper interpretation of $T_p(f)$ in the exceptional case when $f^{\circ p}(0) = 0$ will be discussed later.) We want to apply this operator T_p over and over again to carefully chosen quadratic polynomials.

Conjecture 4.2

If $\mu = s^\infty$ is a generalized Feigenbaum point, where s has period p, then the successive images of the quadratic polynomial q_μ under T_p,

$$T_p^{\circ n} q_\mu(z) = q_\mu^{\circ p^n}(\alpha_n z)/\alpha_n, \qquad \text{where} \quad \alpha_n = q_\mu^{\circ p^n}(0)$$

converge throughout a suitable neighborhood N_0 of the origin to a function F(z) which is a fixed point of the operator T_p and which is quadratic-like. Furthermore, this limit function F is quasi-conformally equivalent to the original polynomial q_μ.

Compare Definition 2.8. (Some authors require that all functions f be normalized so that f(0) = 1. If this convention were adopted, then the quadratic polynomial $q_\mu(z) = z^2 + \mu$ would have to be replaced throughout by the polynomial $q_\mu(\mu z)/\mu = 1 + \mu z^2$. The case $\mu = 0$ must be excluded.)

Thus F must satisfy the Cvitanovic-Feigenbaum functional equation $T_p(F) = F$, or more explicitly:

$$F(z) = F^{\circ p}(\alpha z)/\alpha, \qquad \alpha = F^{\circ p}(0) \tag{4.3}$$

for all $z \in N_0$. Here $|\alpha| < 1$. This convergence of successive images under T_p to a fixed point of T_p has been proved only in a few special cases. (Compare the papers referred to in Section 1.) Note that the entire orbit of 0 under F must lie in the set $K(F) \subset N_0$, while the orbit under $F^{\circ p}$ must lie in the smaller set $\alpha K(F) = K(F^{\circ p} \mid \alpha N_0) \subset K(F)$.

Conjecture 4.4

Conversely, for any $p \geq 2$, any fixed point $F = T_p(F)$ which is quadratic-like with $0 \in K(F)$ arises by this construction from a (necessarily unique) generalized Feigenbaum point s^∞, with s of period p.

Here it is essential to assume that $F : N_0 \to C$ is quadratic-like with $0 \in K(F)$. If, for example, we start with a fixed point $F = T_4(F)$ of T_4 which corresponds to an indecomposable superstable point of period 4, then the image of F under T_2 must also be a fixed point of $T_4 = T_2 \circ T_2$. But this image $G = T_2(F)$ is presumably not quadratic-like with $0 \in K(G)$ and so does not correspond to any point of the Mandelbrot set.

By way of contrast, consider a composite superstable point $s * s'$ of period pp'. In this case one expects that the two generalized Feigenbaum points $(s * s')^\infty$ and $(s' * s)^\infty$ give rise to two distinct fixed points of $T_{pp'}$ which are transformed into each other by T_p and $T_{p'}$, respectively. It is conjectured that these two generalized Feigenbaum points share a common value of the expansion constant δ.

In the cases where 4.2 is known, the proof involves much more precise statements, which are also conjectured to hold in general. Thus, if the operator T_p acts on a carefully chosen space of quadratic-like functions, it is conjectured that F is an isolated fixed point and that the derivative of this operator at F has just one eigenvalue δ of absolute value greater than 1, with all other eigenvalues of absolute value strictly less than 1. This δ is the same expansion constant which occurs in Conjectures 1.1 and 3.1.

By definition, a function f in our space belongs to the stable manifold through the fixed function F if the successive images $T_p^{\circ n}(f)$ converge toward F and belongs to the corresponding unstable manifold if there exists a sequence of functions f_n converging to F so that $T_p^{\circ n}(f_n) = f$. The stable manifold is infinite-dimensional and contains the quadratic map q_{s^∞}, while the unstable manifold is one-dimensional. It has a preferred parametrization which is closely related to the embedding $\psi : M \to C$ of Conjecture 3.3, as follows.

Conjecture 4.5

For any parameter $\tau \neq \psi(0)$ in a neighborhood of the compact set $\psi(M) \subset C$, consider the sequence of points $v(n) = s^\infty + \tau/\delta^n$. If the operator T_p is applied n times to the polynomial $q_{v(n)}$, then as $n \to \infty$ the resulting sequence of functions

$$z \mapsto T_p^{\circ n} q_{v(n)}(z) = q_{v(n)}^{\circ p^n}(\alpha_n z)/\alpha_n, \qquad \alpha_n = q_{v(n)}^{\circ p^n}(0)$$

converges uniformly, throughout a suitable neighborhood N_τ of zero, to a quadratic-like function $z \mapsto \nu_\tau(z)$. In fact, if $\tau = \psi(\mu)$, then ν_τ is quasi-conformally equivalent to the polynomial q_μ. The expression $\nu_\tau(z)$ is analytic as a function of two complex variables.

Note that this parametrization is such that the operator T_p acts linearly,

$$T_p(\nu_\tau) = \nu_{\delta\tau} \tag{4.6}$$

with eigenvalue δ. It follows that each of these functions ν_τ belongs to the unstable manifold through the fixed function $F = \nu_0$. Evidently the study of these unstable manifold functions ν_τ must be closely related to the study of the embedded image $\psi(M)$. For example, one expects that the orbit of zero under ν_τ never escapes from N_τ if and only if $\tau \in \psi(M)$. Similarly, if ν_τ is analytically continued as far as possible, then one expects that the orbit of zero under ν_τ remains bounded if and only if τ belongs to one of the magnified images $\psi(M)\delta^n$.

The special parameter value $\tau = \psi(0)$ is anomalous, since the functions ν_τ converge as $\tau \to \psi(0)$ to a function which is constant (identically equal to one) and hence is not quadratic-like. However, this anomalous behavior is an artifact of our particular choice of normalization and can be eliminated if we are willing to interpret our constructions somewhat differently. We will need the following.

Definition 4.7

Let f and g be analytic functions of one complex variable, each defined and not identically zero in some neighborhood of the origin, and with a critical point at the origin. We will say that f is equivalent to g under rescaling, written $f \cong g$, if there is some constant $c \neq 0$ so that $g(z) = f(cz)/c$ for all z close to zero.

(In most cases, this constant c is uniquely determined.) If $f(0) \neq 0$, note that there is one and only one function $g(z) = f(f(0)z)/f(0)$ which is equivalent to f and which satisfies the normalization condition $g(0) = 1$. However, if f vanishes at the origin, then we must normalize in some other way, for example, by requiring that some higher derivative takes the value one at the origin. There is no method of normalization which works simultaneously for all functions. To see this, it suffices to consider the space of equivalence classes of nonzero polynomials of the form $f(z) = a + bz^2$. This space has the topology of a two-dimensional sphere, and there is a topological obstruction to choosing a representative polynomial in each equivalence class in a continuous manner.

Using this concept of equivalence, the Cvitanovic-Feigenbaum operator satisfies $T_p(f) \cong f^{\circ p}$; the Cvitanovic-Feigenbaum functional equation (4.3) can be written briefly as $F^{\circ p} \cong F$; and equation (4.6) can be rewritten as

$$\nu_\tau^{\circ p} \cong \nu_{\delta\tau} \qquad [4.8]$$

This suggests that we try to work directly with the operator $f \mapsto f^{\circ p}$, which is perfectly well behaved even in cases where $f^{\circ p}(0) = 0$ so that the definition of $T_p(f)$ is problematic. For example, if we believe 4.5, then $\nu_{\psi(s)}$ is a well-defined quadratic-like function. Since $\psi(0) = \delta\psi(s)$ by Section 3, it follows from (4.8) that $\nu_{\psi(0)}$ can be defined, up to equivalence, simply as the p-fold composition $\nu_{\psi(s)}^{\circ p}$. This is a smooth quadratic-like function which vanishes at the origin. Similarly, $\nu_{\delta\psi(0)}$ can be defined, up to equivalence, as the p^2-fold composition $\nu_{\psi(s)}^{\circ p^2}$. This is no longer quadratic-like, but is a smooth function of the form $f(z) = az^{2p}$ + (higher terms).

More generally, we will say that the unstable manifold function ν_τ is defined, at least up to equivalence, whenever there exists some sequence of numbers α_n so that the sequence of rescaled iterates $f_{v(n)}^{\circ p^n}(\alpha_n z)/\alpha_n$ converges, throughout some neighborhood of zero, to a smooth nonconstant limit function. With this form of the definition, ν_τ should be defined and polynomial-like of some degree, in some neighborhood of zero, for a dense open set of parameter values τ. (Compare the discussion in Section 7.)

5. ESCAPE TIME, EXTERNAL ARGUMENTS, AND A DISTANCE ESTIMATE

Douady and Hubbard begin the study of the complement of the Mandelbrot set by observing that there is a conformal mapping

$$\phi : C - M \to C - \bar{D}$$

unique up to a rotation of $C - \overline{D}$, which carries the complement of the Mandelbrot set biholomorphically onto the complement of the closed unit disk. We normalize by requiring that positive real values of μ correspond to positive real values of $\phi(\mu)$, or equivalently by requiring that $\phi(\mu)$ is asymptotic to μ as $|\mu| \to \infty$. The logarithm

$$G(\mu) = \log|\phi(\mu)| \qquad [5.1]$$

will be called the natural potential function on the complement of the Mandelbrot set. It can be characterized as the unique harmonic function on $C - M$ which is asymptotic to $\log|\mu|$ at infinity, and which tends to zero as we approach M.

Closely related is the escape time $\varepsilon(\mu)$ associated with a point $\mu \in C - M$, which we define by the formula

$$\varepsilon(\mu) = -\log_2 G(\mu) \qquad [5.2]$$

This is a smooth function which tends to $+\infty$ as μ approaches M. It should be thought of as a smoothed out approximation to the number of times $n(\mu,B)$ which the function q_μ must be iterated, starting at the origin, in order to escape from some fixed large ball $B \supset M$. In fact, it is not difficult to check that the difference $\varepsilon(\mu) - n(\mu,B)$ remains uniformly bounded as μ tends to M. [Compare (5.7) below.]

A fundamental conjecture of Douady and Hubbard, completely equivalent to 1.5, asserts that: The inverse map ϕ^{-1} extends continuously over the boundary of the unit disk. (They prove the weaker result that ϕ^{-1} tends to a limit as we approach ∂D along an external ray.) If we restrict this extended map

$$\phi^{-1} : C - D \to C - \text{interior}(M)$$

to the boundary of D, then the resulting map

$$\alpha \quad \phi^{-1}(e^{2\pi i\alpha}) \qquad [5.3]$$

from the circle R/Z of reals modulo one onto the boundary of M will be called the natural parametrization of ∂M, and α will be called the external argument of the boundary point $\phi^{-1}(e^{2\pi i\alpha})$. Caution: This function (5.3) is many-to-one, so that a given boundary point of M may well have more than one external argument. As an example, the Misiurewicz point illustrated in Figure 3 has three distinct external arguments.

More generally, for any $\mu \in C$ − interior(M) the external argument $\mathrm{Arg}(\mu) \in R/Z$ is defined by

$$\phi(\mu)/|\phi(\mu)| = e^{2\pi i \mathrm{Arg}(\mu)} \qquad [5.4]$$

This external argument function is well defined and smooth on the complement of M, but is many-valued on the boundary.

Of particular interest are the boundary points with rational external argument, or in other words the images of roots of unity under the map ϕ^{-1}. If α is a rational number of the form $m/(2^p - 1)$, so that the binary expansion of α has period p, then Douady and Hubbard show that $\phi^{-1}(e^{2\pi i\alpha})$ is the root point of a hyperbolic component of period p. Every such root point has exactly two external arguments, both of this form. On the other hand, if α is a rational number with even denominator, so that the binary expansion is eventually periodic, then $\phi^{-1}(e^{2\pi i\alpha})$ is a Misiurewicz point. (If $\mu \in R \cap \partial M$ is any real boundary point, then the bits in the binary expansion of $\mathrm{Arg}(\mu)$ are essentially just the coefficients of the kneading invariant of q_μ, as defined by Milnor and Thurston [18].)

In terms of this external argument parametrization, they give an explicit formula for the product $s * \mu$ with $\mu \in \partial M$, as follows. Let $0 < m/(2^p - 1) < m'/(2^p - 1) < 1$ be the two external arguments at the root point of s, and let

$$m = \sum_{k=1}^{p} m_k 2^{p-k}, \qquad m' = \sum_{k=1}^{p} m'_k 2^{p-k}$$

be the binary expansions of the two numerators.

Assertion 5.5

The binary expansion of $\mathrm{Arg}(s * \mu)$ can be obtained from the binary expansion of $\mathrm{Arg}(\mu)$ by replacing each zero by the sequence $m_1, \ldots, m_p$ and each one by the sequence $m'_1, \ldots, m'_p$.

It follows that $\mathrm{Arg}(s * \mu)$ necessarily belongs to the closed interval $[m/(2^p - 1), m'/(2^p - 1)]$. The correspondence $\mathrm{Arg}(\mu) \longmapsto \mathrm{Arg}(s * \mu)$ is strictly monotone, but has a jump discontinuity at every dyadic rational value of $\mathrm{Arg}(\mu)$ inside the unit interval. (Compare Remark 6.3.)

Suppose that we are given some point μ in the complement of the Mandelbrot set M and want to estimate its distance from M. The following inequality is due to Thurston (unpublished). Let $G'(\mu)$ be the gradient of the potential function G at μ.

Lemma 5.6

The distance of μ from the Mandelbrot set M is strictly less than $2 \sinh G(\mu) / \| G'(\mu) \|$.

<u>Proof</u>. Let $w(\mu) = 1/\phi(\mu)$. Thus w maps the simply connected region $(C \cup \infty) - M$ biholomorphically onto the open unit disk, with $G(\mu) = -\log|w(\mu)|$. Fix some point μ_0 in the complement of M. Let $w_0 = w(\mu_0)$, and let

$$F(w) = (w - w_0)/(1 - w\bar{w}_0)$$

be the fractional linear transformation of the unit disk which takes w_0 to zero, with derivative

$$F'(w_0) = (1 - w_0\bar{w}_0)^{-1} > 0$$

Let r be the distance of μ_0 from M. Then the composition $\mu \mapsto F(w(\mu))$ maps the disk of radius r about μ_0 to the unit disk, with $F(w(\mu_0)) = 0$. Hence the absolute value of the derivative

$$F'(w(\mu_0))w'(\mu_0)$$

of this composition is less than $1/r$ by Schwarz's lemma. Therefore

$$r < (1 - w_0\bar{w}_0)/|w'(\mu_0)|$$

Substituting $|w_0| = \exp(-G(\mu_0))$ and

$$|w'(\mu_0)| = |\exp(-G(\mu_0))| \; \| G'(\mu_0) \|$$

we see easily that $r < 2 \sinh (G)/ \| G' \|$ as asserted.

For small values of G, we may approximate sinh (G) by G. The numbers $G(\mu)$ and $G/\|G'\|$ are both quite easy to compute, as follows. Starting at the point $z_0 = 0$, compute successive values $z_{k+1} = z_k^2 + \mu$, continuing until we obtain some z_n whose absolute value is very large. Then $G(\mu)$ is approximately equal to $\log|z_n|/2^{n-1}$, and the error in this expression decreases very rapidly with increasing n. It follows that

$$\varepsilon(\mu) = -\log_2 G(\mu) \approx n - 1 - \log_2 \log|z_n| \qquad [5.7]$$

Similarly, if we compute $z_n' = dz_n/d\mu$ by the recursion $z_{k+1}' = 2z_k z_k' + 1$ with $z_0' = 0$, then we find that

$$G(\mu)/\|G'(\mu)\| \approx |z_n/z_n'| \log|z_n|$$

where again the approximation is good when $|z_n|$ is large, and the error decreases more than exponentially with increasing n.

The estimate 5.6 is certainly not sharp. For example, as $|\mu| \to \infty$ we have $2 \sinh G(\mu)/\|G'\| \sim |\mu|^2$, although the distance from μ to M is asymptotic to $|\mu|$. (The expression $(1 - e^{-G})/\|G'\|$ would give a much better fit to the distance.) Nevertheless, when $G(\mu)$ is small the estimate 5.6 does seem to give the right order of magnitude. In fact, in this case I conjecture that: The ratio of $2 \sinh G(\mu)/\|G'\|$ to distance always lies between one and four.

It may well happen that $G(\mu)/\|G'(\mu)\|$ is quite small, even though a random search fails to discover any point of the Mandelbrot set which is close to μ. It is natural to ask how one can actually display a point of the Mandelbrot set near μ in this case. Surprisingly enough, one can usually do quite well simply by using Newton's method to solve the equation $G = 0$. It is interesting to note that the ratio $G(\mu)/\|G'(\mu)\|$ is just the absolute value of the change in μ which is prescribed at each step of Newton's method.

6. BEHAVIOR UNDER RESCALING

Now let us consider the behavior of the escape time $\varepsilon(\mu)$ of (5.2) in the neighborhood of a generalized Feigenbaum point s^∞. Suppose that the superstable point s has period p and that the map $\mu \mapsto s * \mu$ has derivative δ^{-1} at s^∞. Let τ be a complex number close to zero. From the description of functions in the unstable manifold, as conjectured in 4.5, we would expect that the orbit of zero under the quadratic map $q_{s^\infty+\tau}$ is qualitatively similar to the orbit of the p-fold iterate of $q_{s^\infty+\tau/\delta}$, as least as long as this orbit remains reasonably close to zero. In particular, we would expect that the escape time $\varepsilon(s^\infty + \tau/\delta)$ is approximately equal to p times the escape time $\varepsilon(s^\infty + \tau)$. In fact, numerical experiments show that this approximation is often quite accurate. If we recall that $\varepsilon(\mu) = -\log_2 G(\mu)$, then it follows that

$$G(s^\infty + \tau/\delta) \approx G(s^\infty + \tau)^p \qquad [6.1]$$

Again, numerical experiments suggest that this approximation is often quite good. However, if we make a corresponding approximation to the distance estimate of Lemma 5.6, then we find an apparent paradox. Let

$$U(\mu) = 2G(\mu)/\|G'(\mu)\| \qquad [6.2]$$

be our approximate upper bound on the distance between μ and the Mandelbrot set. Differentiating (6.1) and substituting in (6.2), we find that

$$U(s^\infty + \tau/\delta) \approx U(s^\infty + \tau)/(|\delta|p)$$

The factor of $|\delta|$ is just what one would have expected, but the factor of p is a surprise. It means that, if we rescale the Mandelbrot set around s^∞, magnifying all distances by δ, we find that the distance of a randomly chosen point from this magnified Mandelbrot set has been decreased by a factor of p. If this estimate is correct, then as we iterate this magnification the magnified images must fill out the plane more and more densely. In fact, computer-drawn pictures seem to indicate that this is indeed the case. However, as indicated in Section 1, the new filaments which are added under each magnification seem to have extremely small measure, so that a randomly chosen point will almost surely miss them.

Remark 6.3. Douady and Hubbard (not yet published) have a different approach to the complicated behavior of these magnified images based on Assertion 5.5. They note that the correspondence

$$\mathrm{Arg}(\mu) \mapsto \mathrm{Arg}(s * \mu)$$

has a jump discontinuity at every dyadic rational value $m/2^n$ of the external argument. (The corresponding points $\mu = \phi^{-1}(e^{2\pi i m/2^n}) \in \partial M$ are characterized by the property that the orbit of zero under q_μ eventually hits the rightmost fixed point of q_μ.) Geometrically, this means that the embedded subset $s * M \subset M$ has an appendage growing out of it at $s * \mu$ for every one of these points μ, which are known to be dense in ∂M. These appendages, at dyadic rational external arguments, are clearly visible in Figure 2. If we iterate this construction, forming $s^{*n} * \mu$ for increasing values of n, then we must keep adding appendages to appendages over and over again, and it seems quite plausible that, in the limit, they must fill up all of the available space, as conjectured in 1.3.

7. THE SCALING INVARIANT η

This last section will describe another approach to the problem of understanding the behavior of the magnified Mandelbrot set near a generalized Feigenbaum point. Let

$$f(z) = a_0 + a_2 z^2 + \text{(higher terms)}$$

be a complex analytic function having a critical point at the origin, $f'(0) = 0$. As in 4.7, we say that f is equivalent to g under rescaling, written $f \cong g$, if there is a constant $c \neq 0$ so that

$$f(z) = g(cz)/c$$

The simplest number associated with f which is invariant under scaling is the quantity

$$\eta(f) = f(0)f''(0) = 2a_0 a_2 \qquad [7.1]$$

As an example, for the quadratic polynomial $q_\mu(z) = z^2 + \mu$ we have $\eta(q_\mu) = 2\mu$. If $\eta(f) \neq 0$, note that f is equivalent to a unique function of the form

$$g(z) = 1 + \eta(f)z^2/2 + \text{(higher terms)}$$

On the other hand, if f has been normalized so that $f''(0) = 1$, then $\eta(f)$ is simply equal to $f(0)$.

Let f^{on} be the nth iterate of f, and let $z_n = f^{on}(0)$, so that $0 \mapsto z_1 \mapsto z_2 \mapsto \cdots$ is the orbit of zero. (If f is an entire function, or if f is polynomial-like with $0 \in K(f)$, then this entire orbit is certainly defined. Otherwise we must assume that n is small enough so that z_n is defined.) The following basic identity is not difficult to verify.

$$\eta(f^{on}) = f'(z_1) \cdots f'(z_{n-1}) z_n f''(0) \qquad [7.2]$$

If f is a quadratic polynomial, so that f" is constant and $f'(z) = zf''$, this formula simplifies to

$$\eta(f^{on}) = f'(z_1) \cdots f'(z_n) \qquad [7.3]$$

Now consider the unstable manifold function ν_τ of Conjecture 4.5. The function $\eta(\nu_\tau)$, or briefly $\eta(\tau)$, is a complex analytic function of just one complex variable. It is therefore somewhat easier to study than the collection of functions ν_τ. This function $\eta(\tau)$ remains smooth and analytic even in a neighborhood of its zeros, which are points of particular interest, although ν_τ is not defined, or at least is defined only up to equivalence, when $\eta(\tau) = 0$. By definition, $\eta(\tau)$ is equal to the limit as $n \to \infty$ of the expression

$$\eta\left(q_{v(n)}^{op^n}\right), \qquad \text{where} \quad v(n) = s^\infty + \tau/\delta^n \qquad [7.4]$$

For reasonable values of n, this can easily be computed using (7.3). If Conjecture 4.5 is true, then $\eta(\tau)$ is always defined and smooth at least for τ in some neighborhood of the compact set $\psi(M)$. In fact, each function $\eta(\delta^n \tau) = \eta(\nu_\tau^{op^n})$ must be smooth throughout some neighborhood of $\psi(M)$. If we also assume Conjecture 1.3, which asserts that the union of the sets $\psi(M)\delta^n$ is everywhere dense, then it follows that: This function $\eta(\tau)$ is defined and smooth throughout a dense open subset of C.

More generally, if α is any positive rational number with denominator a power of p, then an invariant $\eta(\nu_\tau^{o\alpha})$ can be computed by a formula analogous to (7.4). Let us write this invariant briefly as

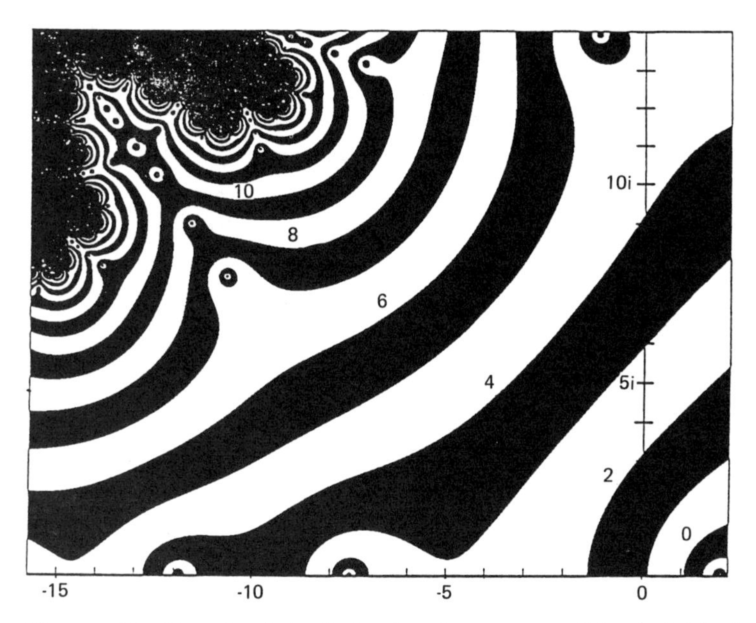

Figure 10 Rough plot of the levels curves of $\log|\eta(\tau)|$, for the function $\eta(\tau)$ associated with the classical Feigenbaum point, as τ varies over a rectangle in the upper half-plane. Here the regions between successive integer values of $\log|\eta|$ are colored alternately black and white. Three zeros of η are visible along the real axis, and a number of zeros clustering toward the chaotic-looking region in the upper left of the picture are also visible. It is conjectured that the function $\eta(\tau)$ is smooth throughout C, except on a closed and nowhere dense singular set which has positive measure in this chaotic-looking region.

$$\eta(\alpha,\tau) = \lim_{n\to\infty} \eta\left(q_{v(n)}^{\alpha p^n}\right), \qquad v(n) = s^\infty + \tau/\delta^n \qquad [7.5]$$

This reduces to the previous expression when $\alpha = 1$. Again it is reasonable to conjecture that this expression is smooth and analytic throughout a dense open set which contains $\psi(M)$. Closely related is the "normalized escape time," which we define by the formula

$$E(\tau) = \lim_{n\to\infty} \varepsilon(s^\infty + \tau/\delta^n)/p^n \leq \infty$$

[Compare (5.2) and (6.1).] It seems plausible to conjecture that this limit exists and that it satisfies $E(\tau) > \alpha$ if and only if $\eta(\alpha,\tau)$ is defined and smooth. However, in view of Conjecture 1.3, we cannot expect the limit function $E(\tau)$ to be continuous, for it must take the value

$+\infty$ on a dense set. The best we can expect is that $E(\tau)$ should be lower semicontinuous.

For the classical case $s^\infty = -1.401155\ldots$, with $\delta = 4.6692\ldots$, the function η, or rather the logarithm of its absolute value, is roughly plotted in Figure 10. This function has Taylor expansion

$$\eta(\tau) = -3.05527 + 2.44113\tau - 0.51330\tau^2 + 0.01613\tau^3 + 0.00848\tau^4 - 0.00144\tau^5 + 0.00009\tau^6 + \cdots$$

The smallest zero, corresponding to the sequence of superstable points s^n of period 2^n converging to s^∞, is at $\tau = \psi(0) = 2.00233\ldots$ and the next smallest zero, corresponding to the sequence of superstable points $s^{*n-2} * (-1.7548\ldots)$ of period $3 \cdot 2^{n-2}$ is at $\delta^2\psi(-1.7548\ldots) = -7.504\ldots$. The radius of convergence of the power series seems to be approximately 15.2.

ACKNOWLEDGMENT

I want to thank E. Bombieri, A. Douady, J.-P. Eckmann, J. Hubbard, O. Lanford, and W. Thurston for extremely useful conversations, and to thank the Institut des Hautes Études Scientifique for its hospitality during the spring of 1985, when much of this work was done.

APPENDIX A: MISIUREWICZ POINTS

Let μ be a point of the Mandelbrot set, and let

$$z_0 = 0 \mapsto z_1 = \mu \mapsto z_2 = \mu^2 + \mu \mapsto z_3 \mapsto \cdots$$

be the orbit of zero under the quadratic map q_μ. By definition, μ is a Misiurewicz point if this sequence $z_0, z_1, z_2, \ldots$ is eventually periodic but not periodic. In other words, the equation $z_n = z_{n+e}$ must be satisfied for some number $e > 0$ and for all large n, but not for $n = 0$. The smallest such number $e > 0$ will be called the "eventual period" of μ, and the smallest t with $z_{t+1} = z_{t+e+1}$ will be called the "transient." Note that $t > 0$ and that t is uniquely characterized by the equation $z_t + z_{t+e} = 0$. Briefly, we will say that μ is a Misiurewicz point of type (t,e).

For any Misiurewicz point μ, Douady and Hubbard show that the periodic orbit $\{z_{t+1}, \ldots, z_{t+e}\}$ is necessarily unstable; that is, the characteristic derivative $\lambda = (q_\mu^{\circ e})'(z_{t+1}) = 2z_{t+1} \cdots 2z_{t+e}$, has absolute value strictly greater than one. As an example, the leftmost point -2 of the Mandelbrot set is the unique Misiurewicz point of type (1,1), with characteristic derivative $\lambda = 4$.

A theorem of Eckmann and Epstein [9], Tan-Lei [19], and Douady and Hubbard asserts that M is self-similar about any Misiurewicz point μ in the following sense: If we translate so that μ goes to the origin and multiply by high powers of λ, then the resulting sequence of magnified images $(M - \mu)\lambda^n$ will converge to a limit in the Hausdorff topology. More precisely, if we compactify by embedding C into the Gauss sphere $C \cup \infty$, then this sequence of compact subsets $(M - \mu)\lambda^n$ will converge to a compact limit set $X \subset C \cup \infty$. Furthermore, according to Tan-Lei, a neighborhood of the origin in X is biholomorphic to a neighborhood of the point μ in the Julia set $K(q_\mu) = \partial K(q_\mu)$. Douady and Hubbard show that this set has measure zero, even though M is equal to the closure of its interior. As an example, if $\mu = -2$ then the limit set X turns out to be the closed interval $[0, \infty]$. Thus the Mandelbrot set has a very sharp point at -2. This is presumably the only example for which the set X is not a fractal.

It is not difficult to see that the set of all Misiurewicz points is everywhere dense in the boundary of M. In fact, every neighborhood N of a boundary point μ_0 must contain some μ such that the orbit of zero under q_μ eventually hits one of the two fixed points $(1 \pm \sqrt{1 - 4\mu})/2$ of q_μ. For otherwise, by Montel's theorem, the family of analytic functions $\mu \mapsto q_\mu^{\circ n}(0)$ on N would be normal. Hence some subsequence would converge to a (possibly infinite) analytic limit function. This is impossible, since we know that this sequence of functions remains bounded at the point μ_0 but diverges to infinity at points arbitrarily close to μ_0.

A similar argument, based on the fact that this family of analytic functions cannot avoid the two values $\pm\sqrt{-\mu}$, shows that every neighborhood of a boundary point contains superstable points. Eckmann and Epstein prove the sharper result that every Misiurewicz point is equal to the limit of a sequence of embedded images $s * M$ of the Mandelbrot set, where the diameters of these embedded images tend to zero very rapidly. (In a highly magnified picture centered at the Misiurewicz point, they are invisible except as centers of configurations which look locally like the set $\pm\sqrt{X}$. These configurations are clearly visible in Figure 3. Since the set X in this case looks roughly like three rays meeting at 120° angles, the corresponding set $\pm\sqrt{X}$ looks like three lines crossing at 60° angles.)

According to Property 2.4, the Douady-Hubbard product operation carries Misiurewicz points to Misiurewicz points. More precisely, using Property 2.9 we see that: If s is superstable of period p, then under the correspondence $\mu \mapsto s * \mu$, the "transient" t of a Misiurewicz point will be multiplied by p, while the eventual period e will be multiplied by some divisor of p. As an example, the Misiurewicz point −2 has type (1,1), while $s(m/p) * (-2)$ has type (p,1). On the other hand, the point $\mu = i$ has type (1,2), while $s(m/p) * i$ has type (p,2p). (Compare Figures 3, 6, and 13.)

APPENDIX B: HUBBARD TREES

The Hubbard tree associated with any superstable point or Misiurewicz point μ in the Mandelbrot set can be described as a caricature of the filled-in Julia set $K(q_\mu)$ of Definition 2.7. It displays the essential information about this set by means of a simplified picture. The constructions given below are slightly different from the original constructions of Hubbard and Douady [5]. We will construct a "reduced Hubbard tree," in which any inessential edges have been collapsed to points. (Another difference from the original construction is that, in the superstable case, we will number edges rather than vertices.) This reduced tree incorporates precisely the minimal information which is needed in order to specify the point μ and to compute products.

First consider a superstable point s of period $p \geq 1$. The reduced Hubbard tree H_s is a connected acyclic graph, with numbered edges $E_1, \ldots, E_p$, which can be constructed as an identification space of $K(q_s)$ as follows. (Compare Figure 11.) Let $0 \mapsto z_1 \mapsto \cdots \vdash z_p = 0$ be the orbit of the critical point under the quadratic map q_s, and let W_j be the component of the interior of $K(q_s)$ which contains z_j. Each closure $\overline{W}_j$ is canonically homeomorphic to the closed unit disk $\overline{D}$. In fact, it

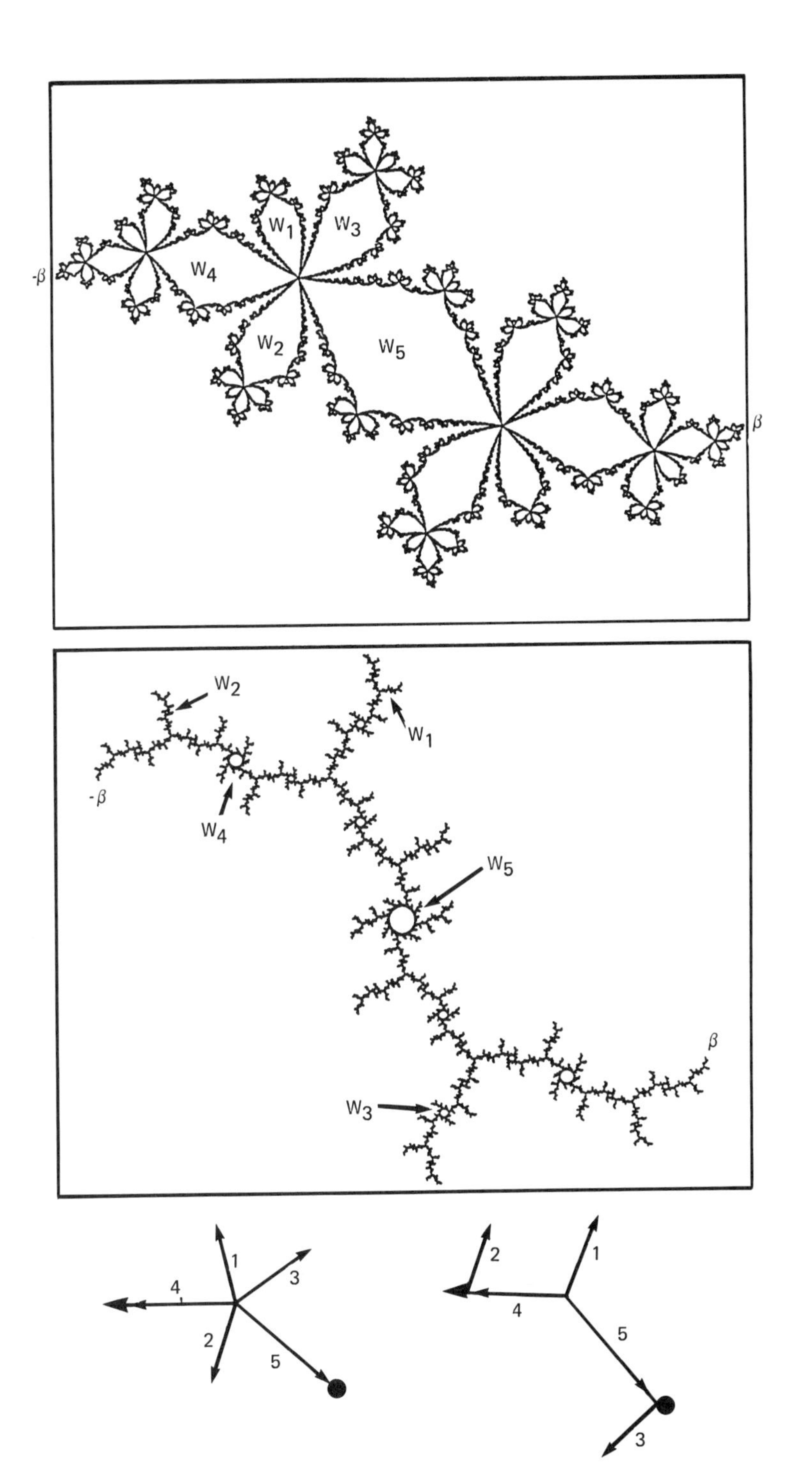

Figure 11 The Julia sets $\partial K(q_s)$ and the reduced Hubbard trees H_s associated with the period 5 satellite $v = s(2/5)$ and the primitive period 5 superstable point $s = -0.442 + 0.9866i$. These trees can be described more compactly by the symbols $[\bar{5},3,\bar{3},1,\bar{1},4;\ \bar{4},2,\bar{2},5]$ and $[\bar{5},1,\bar{1},4,2,\bar{2};\ \bar{4},5,3,\bar{3}]$.

has a preferred local coordinate w, holomorphic on the interior W_j and satisfying $|w| \leq 1$, so that the p-fold iteration of q_s, mapping $\overline{W}_j$ to itself, carries w to w^2. Now form the edge E_j as an identification space of $\overline{W}_j$ by identifying two points if and only if the associated coordinates w have the same real part. Note that each end point of E_j corresponds to a unique point of $\overline{W}_j$, with $w = \pm 1$. The end point with $w = +1$ will be called the "beginning" of the edge E_j or the "root point" of the domain $\overline{W}_j$. In order to paste these edges together into a connected acyclic graph H_s, let us identify an end point of E_j with an end point of E_k whenever these two points either:

1. Represent precisely the same point of $K(q_s)$, or
2. Belong to the closure of a common connected component of the complementary set which is obtained from $K(q_s)$ by removing $\overline{W}_1 \cup \cdots \cup \overline{W}_p$.

(Case 1 occurs if and only if s is a satellite.) It can be shown that there is a canonical continuous map from $K(q_s)$ onto this graph H_s, which collapses each connected component of $K(q_s) - (\overline{W}_1 \cup \cdots \cup \overline{W}_p)$ to a single point. Finally, choose an embedding of H_s into the complex numbers so that the canonical map $K(q_s) \to H_s$ extends to a map from C to itself which induces an orientation-preserving homeomorphism from the complementary set $C - K(q_s)$ onto $C - H_s$. The resulting embedding is well defined up to isotopy.

An argument due to Thurston shows that: This tree H_s, together with the numbering of its edges and the preferred isotopy class of embeddings into C, uniquely determines the superstable point s. In fact, there is an effective algorithm for computing s from H_s.

In order to describe the tree associated with a product $s * s'$, we will need several further pieces of information. First note that each edge E_j of H_s can be given a preferred orientation, which points from "beginning" to end in the direction of decreasing Re(w). Also, we will need two distinguished base points in H_s. Let $\beta \in K(q_s)$ be the rightmost fixed point $(1 + \sqrt{1 - 4s})/2$ of the quadratic map q_s. By the root point for the tree H_s will be meant the image of β under the canonical map from $K(q_s)$ to H_s. Similarly the antiroot point for H_s will mean the image of the negative $-\beta$ under this map. For pictorial purposes, the root point of the tree H_s will be indicated by a heavy dot, and the antiroot point by a solid triangle or arrowhead. If k edges come together at one of these two preferred vertices, with $k \geq 2$, then one further piece of information must also be provided, namely that the distinguished vertex is particularly associated with just one of the k distinct complementary sectors around it. (Compare the second graph in Figure 11.)

As an example, the tree $H_{s(m/p)}$ for the contiguous superstable point s(m/p) of (2.11) consists of p edges pointing out from a common vertex. The angle between consecutively numbered edges is $2\pi m/p$, and the end points of the last two edges are the antiroot point and the root point, respectively (Figures 11 and 13).

The tree associated with a product $s * s'$ can now be constructed simply by replacing each edge E_j of H_s by a copy of $H_{s'}$. The two distinguished vertices of $H_{s'}$ should be placed at the two end points of E_j in such a way that the root point of the tree $H_{s'}$ is inserted at the beginning of the edge E_j. The edges of the resulting product tree are oriented just as in $H_{s'}$ and are to be numbered from 1 to pp' in a modified lexicographical order, so that the copy of edge E'_k which is inserted into E_j is assigned the number $pk + j - p$ in the product tree. The proof of this statement, following Douady and Hubbard, depends on their construction of the pair of topological spaces $K(q_{s*\mu}) \subset C$ from the pair $K(q_s) \subset C$ by removing each component of the interior of $K(q_s)$ and then filling the resulting hole with a copy of $K(q_\mu)$, which is pasted in with suitable boundary identifications.

Just as in Section 5, it is possible to assign "external arguments" to points of the Julia set $\partial K(q_s)$. The two base points β and $-\beta$ have external arguments 0 and 1/2, respectively. Each root point of a component W_j has exactly two external arguments, which are rational numbers of the form (integer)/$(2^p - 1)$ modulo one. These external arguments increase monotonely from zero to one as we traverse the boundary of $K(q_s)$ in a counterclockwise direction. The two external arguments at the root point of W_1 are precisely the same as the two external arguments at the associated root point $s * (1/4)$ in the Mandelbrot set, and the two external arguments associated with each W_{j+1} can be obtained

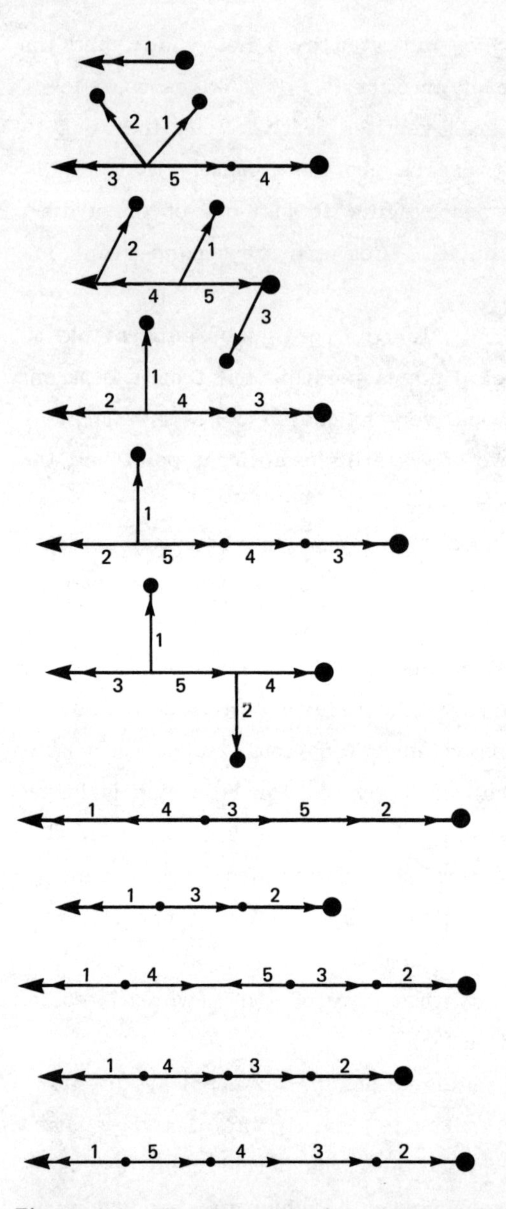

Figure 12 The trees H_s for the primitive superstable points s of period ≤ 5 in the upper half-plane. The identity monoid element s = 0 is listed first, and the others are listed in counterclockwise order around the boundary of M, as in Appendix C.

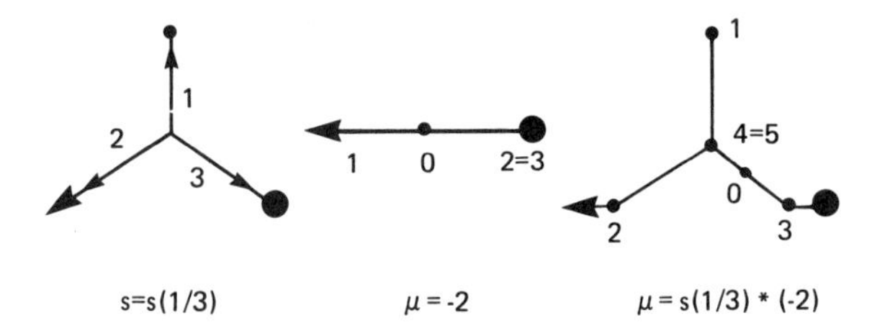

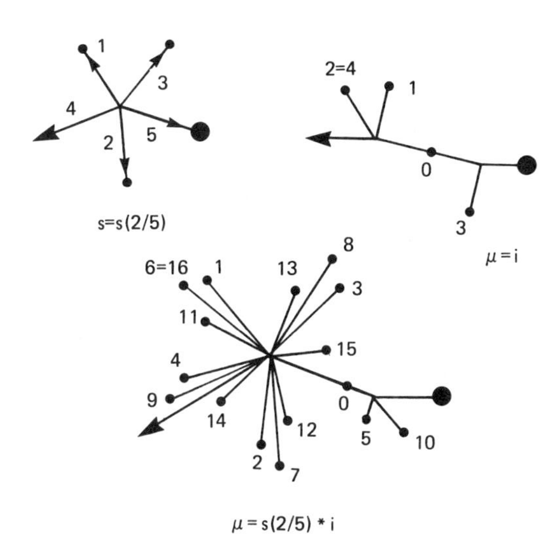

Figure 13 Two examples of product trees $H_{s*\mu}$. Compare Figures 3 and 6.

by doubling those associated with W_j. Using this information, we see that all of these numbers can be effectively computed from the tree H_s, and conversely that the tree H_s is completely determined by the two external arguments at the point $s * (1/4) \in M$.

Now let μ be a Misiurewicz point. As in Appendix A, let $0 = z_0 \mapsto z_1 \mapsto \cdots \mapsto z_{t+e}$ be the orbit of zero under the quadratic map q_μ, with $z_{t+e+1} = z_{t+1}$. We first construct a preliminary version H'_μ of the required tree, as follows. By definition, H'_μ has one numbered vertex v_j for each point z_j in the orbit of zero and one unnumbered vertex for each connected component of the complementary set $K(q_\mu) - \{z_0, z_1, \ldots, z_{t+e}\}$. There is to be an edge joining two vertices if and only if one of the two corresponds to a complementary component and the other corresponds to a point in the closure of this component. There are two distinguished base points in this tree, where the vertex which corresponds to the rightmost fixed point $\beta \in K(q_\mu)$ is called the root point of H'_μ and the vertex which corresponds to $-\beta$ is called the antiroot point. In general, these two base points will be two of the unnumbered vertices; however, in the special case where the orbit of zero ends at the rightmost fixed point β they will coincide with the last two numbered vertices.

The required reduced Hubbard tree H_μ is now obtained from H'_μ by carrying out two minor simplifications, as follows. If a vertex which is neither numbered nor distinguished is incident to just one edge, then we eliminate it by collapsing the edge to a point, while if it is incident to exactly two edges, then we suppress it by fusing the two edges into one longer edge.

As in the superstable case, there is a preferred isotopy class of embeddings of H_μ into the complex numbers. Equivalently, whenever three or more edges come together at a single vertex, there is a preferred cyclic ordering of these incident edges. The Misiurewicz point μ is uniquely determined by this tree H_μ, together with the embedding into C, the numbering of certain vertices, and the information that the $t + e + 1$ − st vertex is to be identified with the $t + 1$ − st.

The product tree $H_{s*\mu}$ can be constructed out of the trees H_s and H_μ in four steps, as follows. (Compare Figure 13.)

Step 1. Enlarge H_s by inserting a new (unnumbered) edge at each of the two distinguished vertices, moving the distinguished vertex out to the end of this edge.

Step 2. Replace part of each numbered edge E_j in the tree H_s by a copy of the Hubbard tree H_μ. This tree should be inserted in place of the middle third of E_j if s is primitive, or in place of the initial half of E_j if s is a satellite. (This distinction is important only in the special case where the orbit of zero under q_μ ends at the rightmost fixed point β.) In either case, the two base points of H_μ should go to the end points of the deleted subinterval, with the root point of H_μ at the end closer to the beginning of E_j.

Step 3. The copy of vertex k which is inserted into E_j is numbered as pk + j − p; however the 0th vertex of H_μ is suppressed in all but the last of the p copies of H_μ, so that no negative numbers occur.

Step 4. Any dangling edges or irrelevant vertices are suppressed, as in the reduction from H'_μ to H_μ above. Also, any edge which joins two unnumbered and undistinguished vertices is collapsed to a point.

APPENDIX C: TABLE OF SUPERSTABLE POINTS

The indecomposable superstable points of period $p \leq 5$ in M can be tabulated as follows. We first list the contiguous superstable points s(m/p) of (2.11) in the upper half-plane, in order of increasing m/p. For more information on these cases, see [13] and also [3,4]. Corresponding to the rough approximation (2.14) there is a very rough approximation to δ, which is listed in the next to last column. The last column lists the interval spanned by the two values of the external argument function of (5.3), at the root point $s * \hat{\mu}(1)$. These values reflect the relative order in which the various hyperbolic components are encountered as we traverse the boundary of M.

m/p	$s = s(m/p)$	s^∞	δ	$\approx -p^2/e^{2\pi i m/p}$	Arg(root)
1/5	.3795136 +.3349323 i	.3770323645 +.3117317763 i	-9.5197 +26.3715 i	-7.7 +23.8 i	[1, 2]/31
1/4	.2822714 +.5300606 i	.30981071737 +.49473250655 i	-.85266 +18.10973 i	16 i	[1, 2]/15
1/3	-.1225612 +.7448618 i	-.0236411685 +.7836606508 i	4.60022 +8.98123 i	4.5 +7.8 i	[1, 2]/7
2/5	-.5043402 +.5627658 i	-.503083185 +.604767482 i	18.96941 +14.56375 i	20.2 +14.7 i	[9,10]/31
1/2	-1.	-1.401155189092	4.669201609103	4	[1, 2]/3

The primitive superstable points with $1 < p \leq 5$ in the upper half-plane are as follows, listed in counterclockwise order around the boundary of M. (Compare Figures 7 and 13.) A very rough approximation to δ is given by the quantity a(s) of (2.13), which is listed in the next to last column.

p	s	s^{∞}	δ	$\approx a(s)$	Arg(root)
5	.3592592 +.6425137 i	.36254287903512 +.64270806921767 i	114.27523 +184.68972 i	108.0 +177.5 i	[3, 4]/31
5	-.0442124 +.9865810 i	-.04475092619563 +.98308682975802 i	-281.883551 -54.458105 i	-279.2 -69.5 i	[5, 6]/31
4	-.1565202 +1.0322471 i	-.15284673572977 +1.03969513783306 i	100.43119 +69.34045 i	96.1 +68.2 i	[3, 4]/15
5	-.1980421 +1.1002695 i	-.19829008064555 +1.10098277848259 i	1399.1592 -253.0024 i	1439.3 -234.5 i	[7, 8]/31
5	-1.2563679 +.3803210 i	-1.25347747897812 +.38268084712978 i	-205.421063 +287.075681 i	-203.8 +280.9 i	[11,12]/31
5	-1.6254137	-1.63192665449011	255.54525	246.4	[13,18]/31
3	-1.7548777	-1.78644025556364	55.24703	52.5	[3, 4]/7
5	-1.8607825	-1.86222402261866	1287.0791187	1281.5	[14,17]/31
4	-1.9407998	-1.94270435069797	981.59498	1008.4	[7, 8]/15
5	-1.9854243	-1.98553953006041	16930.6456	17185.6	[15,16]/31

REFERENCES

1. P. Blanchard, Complex analytic dynamics on the Riemann sphere, Bull. Am. Math. Soc., 11, (1984), 85–141.
2. P. Cvitanovic, Universality in Chaos, Hilger, Bristol (1984).
3. P. Cvitanovic and J. Myrheim, Universality for period n-tuplings in complex mappings, Phys. Lett., 94A, (1983), 329–333.
4. P. Cvitanovic and J. Myrheim, Complex Universality, Nordita (1984).
5. A. Douady, Systemes dynamiques holomorphes, Sem. Bourbaki, 35e annee, #599; Asterisque, 105-6, (1983), 39–63.
6. A. Douady and J. H. Hubbard, Iteration des polynomes quadratiques complexes, C. R. Acad. Sci. Paris, 294, (1982), 123–126.
7. A. Douady and J. H. Hubbard, Etude dynamique des polynomes complexes, Orsay (1984–85).
8. A. Douady and J. H. Hubbard, On the dynamics of polynomial-like mappings, Ann. Sci. École Norm. Sup. Paris, 18, (1985), 287–343.
9. J.-P. Eckmann and H. Epstein, Scaling of Mandelbrot sets generated by critical point preperiodicity, Comm. Math. Phys., 101, (1985), 283–289.
10. J.-P. Eckmann, H. Epstein, and P. Wittwer, Fixed points of Feigenbaum type for the equation $f^p(\lambda x) = \lambda f(x)$, Comm. Math. Phys., 93, (1984), 495–516,

11. M. Feigenbaum, The universal metric properties of nonlinear transformations, J. Stat. Phys., 21, (1979), 669–706. (Reprinted in [2].)

12. M. Feigenbaum, Universal behavior in nonlinear systems, Los Alamos Sci., 1, (1980), 4–27. (Reprinted in [2].)

13. A. I. Golberg, Ya. G. Sinai and K. M. Khanin, Universal properties of a sequence of period tripling bifurcations, Usp. Mat. N., 38, (1983), 159–160. (Russ. Math. Surv. 38, (1983), 187–188.)

14. O. Lanford, A computer-assisted proof of the Feigenbaum conjecture, Bull. Am. Math. Soc., 6, (1982), 427–434. (Reprinted in [2].)

15. S. Levy, Critically finite rational maps, Thesis, Princeton University (1985).

16. B. Mandelbrot, The Fractal Geometry of Nature, Freeman, San Francisco (1982).

17. B. Mandelbrot, On the dynamics of iterated maps III, "The individual molecules of the M-set, self-similar properties, the empirical n^2 rule, and the n^2 conjecture," and V, "Conjecture that the boundary of the M-set has a fractal dimension equal to 2," pp. 213–224 and 235–238 of Chaos, Fractals, and Dynamics, P. Fischer and W. Smith, eds., Marcel Dekker, New York (1985).

18. J. Milnor and W. Thurston, Iterated maps of the interval (to appear).

19. Tan-Lei, Ressemblance entre l'ensemble de Mandelbrot et l'ensemble de Julia au voisinage d'un point de Misiurewicz (Section 23 of [7] above).

20. W. Thurston, On the combinatorics of iterated rational maps (in preparation).

12

Homotopy Groups of Spheres on a Small Computer

DOUGLAS C. RAVENEL

University of Rochester
Rochester, New York

This chapter is a preliminary report on a program designed to compute homotopy groups of spheres using the EHP sequence on a 64K computer. The determination of these groups is one of the oldest and most difficult computational problems in algebraic topology. For mathematical background we refer the interested reader to [5].

Very briefly, we are interested in finding abelian groups $\pi_{n+k}(S^n)$, defined to be the set of continuous maps from an $(n + k)$-dimensional sphere to an n-dimensional sphere, classified up to continuous deformation. These are known to vanish for $k < 0$ and to be finite unless $k = 0$ or $k = n - 1$. They are also known to be independent of n if $n > k + 1$. Thus for a given value of k there is only a finite number of these groups to compute. The method we will use involves induction on k.

Since these groups are abelian and (in nearly all cases) finite, we can ask for their p-primary component for each prime number p. These p-local calculations can be done independently of each other. Most research in this area has concentrated on prime 2, but we have reason to believe that the odd primary case is easier and in some sense more informative. Our program is designed to work equally well at any odd prime, and the first thing it does is ask which prime the user wants to work with. The actual calculations we have made with it are at the primes 3 and 5.

Previous attempts [1,7] at such computations have been made on far more powerful machines and have been very costly in terms of storage and CPU time. Moreover, the output of these programs has not been the homotopy groups themselves, but the E_2-term of the unstable Adams spectral sequence. This is a certain algebraic approximation to the desired answer. Additional calculations (the differentials in the Adams spectral sequence), outside the scope of these programs, are required to get the actual homotopy groups. For small values of k this extra work is minimal, but it is clear that it will become overwhelming as k gets larger.

More precisely, for small k most elements in the Adams E_2-term are permanent cycles, but as k grows the proportion of elements that support nontrivial differentials approaches unity. (For $p = 3$ this is illustrated below in Table 3.) There is no published theorem to this effect, nor is there likely to be in the foreseeable future. The skeptical reader may regard it as an unproved conjecture. It is one of many assertions about the Adams spectral sequence that can be made with great confidence by those who have worked with it extensively, but which are not worth the trouble of proving rigorously.

For those who are familiar with the Adams spectral sequence at the prime 2, we can offer the following supporting evidence. The E_2-term has a vanishing line of slope 1/2 and a lot of elements (including the Mahowald-Tangora wedge [4]) above a line of slope 1/5. However, the results of [3] indicate that in the E_2-term all that remain in this region are a few elements near the vanishing line related to the image of the J homomorphism. Thus the set of elements that survive to E_2 in this region can be said to have measure zero.

Long, hard experience with this problem has led to the following.

Mahowald Uncertainty Principle:

1. Any spectral sequence (or other algebraic construction) converging to the homotopy groups of spheres with a systematically computable E_2-term (or other input) will have differentials (or other extra work required) that can be computed only by ad hoc methods.

2. Any computer program that will yield interesting information about homotopy groups of spheres will require repeated human interaction while it is running.

The calculations we have in mind could not possibly be performed entirely by any program alone. Our strategy is to use the computer to do all of the necessary bookkeeping and the easier parts of the calculation, leaving the more difficult decisions to the intelligent user. The nature of the computation is such that one's understanding of it evolves as one does it. Thus one is inclined to make changes in the program along the way. The main idea of this chapter is a format (more specifically a data structure suitable for storing the requisite information) within which such modifications can be made without having to redo the program and all the calculations from scratch each time one discovers a new shortcut.

One of our guiding principles in this endeavor is the following.

Tangora's Law

Computation precedes theory or "answers now, proofs later."

This means that we do not insist on complete mathematical rigor at every stage of the program development. If we find a computational shortcut that appears to be correct most of the time, then we put it into the program on an experimental basis. Experience has shown that mistakes in this type of calculation quickly lead to obvious contradictions. Moreover, this calculation is so complicated that even partial information is useful.

In Section 1 we describe mathematical problems more precisely along with the simplest form of the program. In Sections 2 and 3 we describe some mathematical and computational refinements that make the calculation go faster. In Section 4 we describe some empirical observations about homotopy groups of spheres. We are not including a detailed listing of the program or its output as we hope to improve on both in the near future.

1. THE EHP SEQUENCE AND THE FIRST DRAFT OF THE PROGRAM

At the prime 2 one has fibrations

$$S^n \to \Omega S^{n+1} \to \Omega S^{2n+1}$$

There are similar fibrations at odd primes which entail more complicated notation; see (1.7) below. The resulting long exact sequence of homotopy groups is called the EHP sequence. It has the form

$$\to \pi_{n+k}(S^n) \overset{E}{\to} \pi_{n+k+1}(\Omega S^{n+1}) \overset{H}{\to} \pi_{n+k+1}(\Omega S^{2n+1}) \overset{P}{\to} \cdots \qquad [1.1]$$

The map H is called the Hopf invariant.

Combining these for all n leads to an exact couple and a spectral sequence known as the EHP spectral sequence, which we will abbreviate by EHPSS. Its E_1-term consists of the homotopy groups (localized at the prime p) of all the odd-dimensional spheres, and its E_2-term consists of the stable homotopy groups of spheres $\pi_k(S^0)$ is filtered by the images of the unstable groups $\pi_{n+k}(S^n)$ for various n. We call this the sphere of origin filtration.

The EHPSS can easily be truncated in such a way as to converge instead to $\pi_*(S^m)$ for any m. The groups needed as input in the E_1-term to compute the k-stem are in lower stems which have been computed inductively. If information about the lower stems is stored in memory, then it is easy to write a subroutine (Section 1.5) that searches for the input needed for the next stem. One then needs somehow to compute and record in memory (Section 1.4) the differentials going from the k-stem to the (k − 1)-stem.

Before describing our data structure and program in more detail we need to discuss some general properties of the EHPSS. The E_1-term is a bigraded abelian p-group where each bigrading is a previously computed homotopy group of some odd-dimensional sphere. These groups are known to be finite in positive stems. A special provision in the program must be made to handle the infinite group

$$\pi_{2n+1}(S^{2n+1}) = Z$$

which we discuss below.

Each such finite group is filtered by sphere of origin just as the stable groups are. The subquotients in this filtration are themselves filtered in the same way. Iterating this filtration, we arrive at one in which each nontrivial subquotient is Z/p. Thus we get

Proposition 1.2

The E_1-term of the EHPSS is a Z/p-vector space of finite type with a canonical basis.

Each differential amounts to a map between one-dimensional vector spaces, and all that matters is whether it is zero or not. We do not need a matrix to describe such a differential.

Moreover, each basis element has a Hopf invariant which is another basis element, so the canonical basis has a tree structure.

Now we are ready to describe our method of storing information for the EHPSS.

1.3 Data Structure

For each basis element X we create a record consisting of the stem ST(X) of X and four pointers: HI(X), TG(X), PM(X), and SH(X). These are addresses of previously defined basis elements. HI(X) is the address of the Hopf invariant of X. The other three are specified below. They are given a default value when the record is first created.

On a 64K computer we allocate 10 bytes for each basis element, 2 for the stem, and 2 for each of the pointers. For the primes 2 and 3, two bytes for the stem may seem excessive since we are not likely to reach the 256-stem, but for the prime 5 this dimension is quite accessible.

The basis elements are created and stored stem by stem, within each stem by sphere of origin, and within each sphere of origin by the order in which their Hopf invariants were stored. In a separate area of memory we store the address of the first basis element in each stem. We call these addresses stem pointers. This information is helpful in the search routine used to create the basis element for the next stem.

The sphere of origin of an element X can easily be determined from the information stored by comparing the stems of X and its Hopf invariant. Now suppose X is in the k-stem and we have somehow determined that $d_r(X) = Y$, where Y is a basis element in the (k − 1)-stem. The index r means that the sphere of origin of X exceeds that of Y by r. This means that Y corresponds to an unstable homotopy element, i.e., one that dies (becomes null homotopic) after r suspensions. If S^m is the sphere of origin of X, we say that S^m is the sphere of death of Y.

1.4 Method of Recording Differentials

If $d_r(X) = Y$ as above, then we set TG(Y) equal to HI(X) and delete the record of X from memory.

TG is short for "tag," the term for this element coined by Tangora. The sphere of death of Y is easily computed by comparing the stems of Y and TG(Y). The element X is removed because we have no further use for it. Since it is not a cycle in the E_r-term of the EHPSS, it does not survive to the E_{r+1}-term. Experience shows that in the k-stem, the proportion of elements which support nontrivial differentials is roughly (k − 1)/2k, so removal of these elements saves a great deal of storage space.

Now we will describe in detail our inductive method for creating the basis for the k-stem. At the start of the program we create an element in the 0-stem whose Hopf invariant is itself. This is the fundamental class i on one.

For the prime 2 we have

$$E_1^{n,k} = \pi_{n+k}(S^{2n-1})$$

If an element in this group is a permanent cycle in the EHPSS, then it corresponds to a homotopy element born on S^n. For an odd prime the indices need to be altered appropriately. We need to look up these previously calculated homotopy groups for $2 \le n \le k - 1$. For $n < k - 1$ we proceed as follows.

1.5 Method for Creating $E_1^{n,k}$

Using the stem pointer, we begin looking through the (k − n + 1)-stem. As soon as we come to an element whose sphere of origin exceeds 2n −

1, or whose stem exceeds $n - k + 1$, we move on to the next value of n. If an element X is unstable (that is, if TG(X) is not the default value) and its sphere of death is less than $2n - 1$, we ignore it. Otherwise we create a new record Y with ST(Y) = k and HI(Y) equal to the address of X. Then we move on to the next element in the $(n - k + 1)$-stem.

Elements in the $(n - k + 1)$-stem born after or dying before S^{2n-1} are irrelevant because they do not contribute to the relevant homotopy group.

When we get up to $n = k - 1$ we have to proceed differently. The group in question is the integers Z. Denote its generator by x_k. Then we know that

$$d_1(x_k) = 2x_{k-1} \qquad \text{for } k \text{ even}$$

so

$$E_2^{k-1,k} = Z/2 \qquad \text{if } k \text{ is odd}$$
$$= 0 \qquad \text{if } k \text{ is even}$$

We cannot store the infinite group $E_1^{k-1,k}$ so we store $E_2^{k-1,k}$ instead. In other words, if k is odd we create an element with Hopf invariant one, and if n is even we create no additional element. This means that the output of the program will not reflect the infinite summand of $\pi_{4m-1}(S^{2m})$.

Now we can describe the main program loop in its simplest form (compare with Definition 3.12). Assume inductively that we have created all the basis elements in stems less than k and that we have computed all the differentials originating in those stems. The records for stems less than $k - 1$ have been printed (on paper) on a table. (See sample below.)

1.6 Main Program Loop

Step A. Create basis in k-stem as described in Section 1.5.

Step B. See if any differentials are logically possible. If none are then go to D. If there could be a differential then display the (k − 1)- and k-stems on the screen and ask the user if there are any more. If the answer is yes then go on to C; otherwise go to D.

Step C. Ask the user to specify the source and target of the next differential, which the user must compute by hand. Record it as in Section 1.4. Go to B.

Step D. Add the (k − 1)-stem to the printed table.

Stem E. Increase k by one and go to A.

In step B the program looks to see if both the k-stem is nonempty and the (k − 1)-stem still has elements which have not been tagged. If both of these conditions are met then it looks to see if the youngest element in the k-stem (i.e., the one born on the biggest sphere, the last one on the list) is younger than the oldest untagged element in the (k − 1)-stem. We go to step C only if all these conditions are met.

Before giving a sample table for $p = 3$, we need to spell out the modifications needed for odd primes. The fibration of (1.1) is replaced by a pair of fibrations.

$$X^{2m} \to \Omega S^{2m+1} \to \Omega S^{2pm+1}$$

and [1.7]

$$S^{2m-1} \to \Omega X^{2m} \to \Omega S^{2pm-1}$$

The space X^{2m} is a substitute for S^{2m} and is the $(2pm - 1)$-skeleton of ΩS^{2m+1}. It is a CW-complex with one cell in each of the first $p - 1$ dimensions divisible by 2m. The homotopy groups of S^{2m} itself are determined by the odd primary equivalence (due to Serre)

$$\Omega S^{2m} = S^{2m-1} \times \Omega S^{4m-1}$$

In the resulting EHPSS one has

$$E_1^{2m+1,k} = \pi_{2m+1+k}(S^{2pm+1})$$

and [1.8]

$$E_1^{2m,k} = \pi_{2m+k}(S^{2pm-1})$$

1.9 Sample Printed Table for p = 3

Address	Birth		Death	
0-stem				
0	0	0		
3-stem				
1	2	0		
6-stem				
2	2	1	4	0
7-stem				
3	3	1		
10-stem				
4	2	3	6	0
5	4	1		
11-stem				
6	3	3		
7	5	1		

In the printed table the basis elements are numbered, starting with 0 for ι, and these numbers are treated as addresses. In the Birth column are two numbers: the sphere of origin and the address of the Hopf invariant. In the Death column, which is left blank if the element is stable, the two numbers are sphere of death and the address of the tag.

The five stable elements shown in positive stems in the table are respectively α_1, α_2, β_1, α_3 and α_3'.

This table does not show, nor does the program keep track of, group extensions. Nontrivial extensions do occur in the 10- and 11-stems. Knowing this, the table gives the following 3-primary homotopy groups in the 10-stem.

$$\pi_{12}(S^2) = \pi_{13}(S^3) = Z/3$$

$$\pi_{14}(X^4) = \pi_{15}(S^5) = Z/9$$

$$\pi_{16}(X^6) = \pi_{10+n}(S^n) = Z/3 \qquad \text{for } n > 6$$

The table that appears on the screen after the 11-stem has been created as follows.

1.10 Sample Screen Table for p = 3

Address	Birth		Death
10-stem			
4	2	3	
5	4	1	
11-stem			
6	3	3	
7	5	1	
8	6	0	

Notice that the last element here is not present in 1.9. The program asks for a differential, and we tell it that there is one whose source is 8 and whose target is 4. Thus element 4 gets a tag value of 0 (the Hopf invariant of 8) and element 8 is removed as described in Section 1.4, and the table changes to what we see in Section 1.9.

2. EXCLUDING THE IMAGE OF J FROM THE CALCULATION

In this we give a method of taking advantage of known results which describe a large portion of the canonical basis systematically. In low dimensions this portion is more than half. We will be more precise about this in Section 4.

The theorems that we wish to exploit are due to Gray [2], Mahowald [3], and Thompson [8]. They concern elements whose Hopf invariants lie on the image of the J homomorphism. For odd primes this image consists of the α family. For more on this we refer the reader to [5,6].

The results in question describe a large portion of the EHPSS. It is desirable to exclude this portion from our calculation since it is already well understood. We do this by modifying the process by which the basis is constructed in Section 1.5.

Mathematically, the EHPSS is an inductive process in which the induction is started by our knowledge of $\pi_*(S^1)$. Computationally, we have to alter this slightly since our program cannot deal with infinite groups. For p = 2 we insert an element with Hopf invariant one in every odd stem. For an odd prime p we do this in each stem congruent to $-1 \bmod 2p - 2$. These elements, rather than $\pi_*(S^1)$, are the effective inductive input.

We will call this collection of elements, along with ι, the input file. We include i so that the input file will be closed under the Hopf invariant, i.e., so that it will have a tree structure. Basis elements in it are stored with the same data structure (Section 1.3) that is used in the main program.

We will exclude Im J from the program by using a different input file, to be described below. It is more complicated than the one used in the previous section. It requires a mathematical justification which is beyond the scope of the chapter and which will be given in [6].

This input file is actually created in a separate program and stored on a disk. The main program begins by asking for the name of this input file, which it loads into memory. This input file determines the prime and the top stem the program could reach if it were to run long enough.

It has a tree structure, and the elements which are to be inserted in 1.5 are precisely the ends of the tree, i.e., the elements which are not Hopf invariants of other elements in the input file. These elements are listed at the end of the file in the usual order. The file includes a pointer to the first of these elements, and the program keeps track of the stem and sphere of origin of the next one to be inserted in the basis.

The following table shows the input file that would be used at the prime 3 if we wished to compute up to the 50-stem. The addresses shown are negative because they are relative to the address of the first basis element. As in Sections 1.9 and 1.10, we list the sphere of origin for convenience even though it is not explicitly stored in memory. The other three pointers, TG, PM, and SH (the last two of which we have still not defined), are left at default value and so are not shown in the table.

2.1 Sample Input File for p = 3

Address	Stem	Birth	
−15	0	0	−15
−14	3	2	−15
−13	7	3	−14
−12	11	5	−14
−11	10	2	−13
−10	22	8	−13
− 9	34	12	−12
	* * * *		
− 8	10	4	−14
− 7	13	2	−11
− 6	22	10	−14
− 5	29	4	−10
− 4	34	14	−13
− 3	34	16	−14
− 2	45	6	− 9
− 1	46	22	−14

The first element on the list, −15, is ι, the root of the tree. The next three are the generators of Im J in the indicated stems, namely α_1. α_2, and α_3'.

The next three after them, −11 to −9, are unstable elements, namely Whitehead products. This means that in the usual EHPSS they would be tagged by elements with Hopf invariant one. Mathematically these elements are described by the odd primary vector field theorem (see [5]).

The last eight elements are the ends of the tree, the ones that are to be inserted into the basis for the EHPSS. They fall into two groups according to the parity of the stem.

The even-dimensional elements occur in stems congruent to −2 mod 12 and have Hopf invariants which are stable and in Im J. There is an element with Hopf invariant α_1 in every such stem, one with Hopf invariant α_2 in each stem congruent to −2 mod 36, one with Hopf invariant α_3' in each stem congruent to -2 mod 108, and so on. Thus there would be two such elements in the 34- and 60-stems and three in the 106-stem.

These elements correspond to those in $J_*(B\Sigma_p)$ which are in the kernel of the Kahn-Priddy map to $J_*(S^0)$.

There is an odd-dimensional element in each stem of the form $16m - 3$. It is born on the 2m-sphere and its Hopf invariant is the Whitehead product in the $(12m - 2)$-stem.

More generally, for an odd prime p let $q = 2p - 2$ and let x_i and α_i denote respectively the generator and element of order p in the image of the J-homomorphism in the $(qi - 1)$-stem. These are equal if p does not divide i. It is known that α_1 has Hopf invariant one, α_i has Hopf invariant α_{i-1} for $i > 1$, and x_i has Hopf invariant α_{i-k-1} when i is a unit (in the integers localized at p) multiple of p^k.

The general description of the input file for an odd prime p and an arbitrary range of dimensions is as follows.

Theorem 2.2

With notation as above, the input file consists of the following five families of elements.

1. The fundamental class ι, whose Hopf invariant is itself.
2. The elements x_i and α_i with Hopf invariants as described above. The largest relevant value of i grows logarithmically with the range of dimensions being considered.
3. The Whitehead product w_m in the $(pqm - 2)$-stem for all m such that $(p + 1)qm - 3$ does not exceed our range of dimensions. The Hopf invariant of w_m is x_{k+2}, where p^k is the largest power of p dividing m.
4. The element in the $(p^k mp - 2)$-stem having Hopf invariant x_k for all m and k such that this stem is in the range being considered.
5. The element in the $((p + 1)mq - 3)$-stem with Hopf invariant w_m for all m as in 3.

Elements in the last two families are to be inserted in the EHPSS basis.

3. AUTOMATIC COMPUTATION OF EASY DIFFERENTIALS

After making some computations using the input file described in the previous section we found that most differentials are so easy to predict that they could be computed automatically. We begin this discussion with an illustration. The negative addresses shown as Hopf invariants in the following table refer to those in the input file of Section 2.1.

3.1 Sample Output for p = 3

Address	Birth		Death	
10-stem				
0	4	−14		
13-stem				
1	2	−11	3	0
2	2	0		
16-stem				
3	2	2	4	0
17-stem				
4	3	2	5	0
20-stem				
5	4	2		
21-stem				
6	5	2	7	0
7	6	0	10	−14
23-stem				
8	2	5		
24-stem				
9	2	6	3	7
10	3	5	7	2
11	6	2	8	0
26-stem				
12	2	8	4	5
13	9	0		

Address	Birth		Death	
27-stem				
14	2	10	3	11
15	3	8	5	5
28-stem				
16	8	2	10	0
29-stem				
17	10	−10		
18	9	2	11	0
30-stem				
19	4	8		
31-stem				
20	5	8	7	5
21	6	5	10	2
32-stem				
22	2	17	4	13
33-stem				
23	2	19	14	− 4
24	3	17	5	13
25	11	2	13	0
26	12	0	16	− 3
34-stem				
27	2	20	3	21
28	3	19	7	8
29	6	8	8	5

The stable elements shown here are $\beta_1(0)$, $\alpha_1\beta_1(2)$, $\beta_1^2(5)$, $\alpha_1\beta_1^2(8)$, $\beta_2(13)$, $\alpha_1\beta_2(17)$, and $\beta_1^3(19)$.

The simplest differentials shown in this table are those with length one. An element born on an even-dimensional "sphere" dies after a single suspension. This happens with elements 1, 9, 11, 14, and 27. We call such elements ephemeral. Examination of the fibrations (1.7) shows that such differentials are induced by maps

$$\Omega^3 S^{2pm+1} \to \Omega S^{2pm-1} \tag{3.2}$$

having degree p on the bottom cell.

Definition 3.3

We use the pointer PM to describe the map (3.2). Consequently, we have

$$d_1(X) = Y \quad \text{if } HI(Y) = PM(HI(X))$$

The program asks the user to specify PM(X) after all differentials in the stem of X have been computed.

In our example we would have

$$\begin{aligned} PM(0) &= PM(-8) = -11 \quad \text{(since 0 and } -8 \text{ are actually the same element)} \\ PM(2) &= 1 \\ PM(7) &= 6 \\ PM(11) &= 10 \\ PM(21) &= 20 \\ PM(26) &= 25 \\ PM(28) &= 29 \end{aligned} \tag{3.4}$$

Thus the computer can use the pointer PM to compute d_1 automatically, but that is not its only use.

Proposition 3.5

If X and PM(X) are both alive on an odd-dimensional sphere, then PM(X) is p times X.

To prove this, note that composing the map (3.2) on either side with the double suspension

$$\Omega S^{2pm-1} \to \Omega^3 S^{2pm+1}$$

gives the pth power map, which induces multiplication by p in homotopy.

Corollary 3.6

If X and Y are in the same stem and born on the same sphere and PM(HI(X)) = HI(Y) then Y is p times X.

Thus (3.4) gives us nontrivial group extensions in the 13-stem, but not in stems 21, 24, 27, 31, 33, and 34, since in those cases the two elements in question are not both alive on an odd sphere.

Unfortunately, not all group extensions can be accounted for in this way. For example, in the 33-stem it is known that on S^{13}, 23 is p times 26, but the pointer PM will not tell us that since PM(26) = 25.

The pointer PM itself can be found automatically in many cases. Examination of 3.4 reveals the following.

Experimental fact 3.7. If X and Y are in the same stem and HI(X) = HI(HI(Y)), then PM(X) = Y.

The program can check this condition very easily. In the computations we have done this accounts for nearly all the values of PM. For p = 3 the first "exotic group extension," i.e., the first value of PM not given by 3.7, occurs in the 45-stem. After computing the values of PM given by 3.7 in a given stem, the program asks the user if there are any others.

We now turn to the differentials in Section 3.1 of lengths 2 and 4. The only remaining differential after these is the one of length 12 which kills 23 ($\alpha_1\beta_1^3$) in the 33-stem. This corresponds to the Toda differential in the Adams or Adams-Novikov spectral sequence. It is one of the things which seems to be beyond the reach of any foreseeable computer program.

Examination of Section 3.1 shows:

Experimental fact 3.8. In most cases if r is even and less than 2p, $d_r(X) = Y$ whenever HI(X) = HI(HI(Y)).

The "if and only if" statement that one might make about this has counterexamples on both counts. The differentials which tag elements 22 and 24 do not satisfy the Hopf invariant condition, while elements 5, 8, and 19 are not tagged by the expected differential.

We will deal with the latter phenomenon first. The expected differential does not occur when a certain binomial coefficient vanishes mod p, namely

$$\binom{k}{r/2} \tag{3.9}$$

where

$$k = [(ST(Y) - ST(HI(Y))/2]$$

Conjecture 3.10

Under suitable hypotheses, the Hopf invariant condition of 3.8 and the nonvanishing mod p of the coefficient (3.9) imply the differential of 3.8.

We outline a possible proof. The argument uses the connection between the EHPSS and the Atiyah-Hirzebruch spectral sequence for the stable homotopy of $B\Sigma_p$ (see [5]). There are Dyer-Lashof maps from skeleta of $B\Sigma_p$ to the spaces

$$\Omega^{2n}X^{2n} \quad \text{and} \quad \Omega^{2n+1}S^{2n+1}$$

and Snaith maps from these spaces to QX, where X is a skeleton of $B\Sigma_p$. In many cases an EHP differential is connected to some homotopy relation in $B\Sigma_p$.

The Hopf invariant condition of 3.8 implies that HI(Y) is born on S^{2r}. This usually implies a Toda bracket relation

$$HI(Y) \in \langle \alpha_1, \alpha_1, \ldots, HI(X) \rangle \tag{3.11}$$

with α_1 occurring r/2 times. This gives a d_r in the Atiyah-Hirzebruch spectral sequence for the stable homotopy of $B\Sigma_p$ if the Steenrod operation P^r is nonzero on the appropriate cohomology class, which happens iff the coefficient (3.9) is nonzero.

The hypotheses of 3.10 are satisfied in the metastable case, i.e., when the stem is less than roughly p^2 times the sphere of origin of Y. Moreover, they are satisfied so often that we believe that on a small computer it is not worthwhile to have the program check them each time. Instead, we make a provision for users to veto any such differential they are not sure about.

We now turn to the other type of counterexample to 3.8, namely cases where a differential occurs even though the Hopf invariant condition is not satisfied. This happens with elements 22 and 24 in Section

3.1. In both cases HI(X) is 13 while HI(Y) is 17. These elements are β_2 and $\alpha_1\beta_2$, respectively. In general, if x is a homotopy element presently on S^{s2p-1}, then it is the Hopf invariant of $\alpha_1 x$, but β_2 is not born until S^9, so it cannot be the Hopf invariant of $\alpha_1\beta_2$. Instead the latter is the Whitehead product w_2 of Theorem 2.2(3).

If the Hopf invariant of $\alpha_1\beta_2$ were β_2, then 3.10 would give the two differentials. In other words, $\alpha_1\beta_2$ behaves in the EHPSS as if its Hopf invariant were β_2. We use the pointer SH (for stable Hopf invariant) to record this fact, setting SH(17) = 13 in Section 3.1.

Definition 3.12

The pointer SH, the stable Hopf invariant, is set equal to the Hopf invariant HI unless the user specifies otherwise. The Hopf invariant condition of 3.8 and 3.10 is replaced by HI(X) = SH(HI(Y)).

We have described methods for calculating d_1 and d_r for even $r < 2p$. They involve additional pointers PM and SH, respectively. These are assigned certain default values which the user is free to change. The program asks for these changes after all differentials to and from the stem in question have been determined.

PM(X) and SH(X) are used to compute differentials involving elements having Hopf invariant X. Therefore they are needed only if X can be a Hopf invariant, i.e., only if X is alive on a (2mp − 1)-sphere or a (2mp + 1)-sphere for some m. If X is alive on such a sphere we say it is fertile; otherwise we say it is sterile. For example, ephemeral elements (those born on an X^{2n} and dying on S^{2n+1}) are sterile.

If a stem is found to have no fertile elements, then the program does not ask for exotic values of PM and SH.

Once the easy differentials from the k stem to the (k − 1)-stem have been computed, the program checks (as in step B in Section 1.6) to see if more differentials are logically possible. If they are, the user must calculate them by hand. However, if we knew that an element in the k-stem was the target of an easy differential originating in the (k + 1)-stem, then it would have to be a permanent cycle in the EHPSS.

This information could simplify, and in some cases eliminate, the manual calculation of hard differentials. This suggests reorganizing the program so that it creates the (k + 1)-stem and computes easy differentials from it to the k-stem before asking the user to calculate hard differentials from the k-stem to the (k − 1)-stem. This brings us to the following.

3.13 Improved Main Program Loop

Assume inductively that we have created the k-stem and computed the easy differentials from it to the (k − 1)-stem. The (k − 2)-stem has been printed on paper.

Step A. Create basis for the (k + 1)-stem as in 1.5, using the input file as described in Section 2.

Step B. Compute easy differentials from the (k + 1)-stem to the k-stem, using 3.3 and 3.11.

Step C. See if any differentials from the k-stem to the (k − 1)-stem are logically possible, using the fact that the target of an easy differential must be a permanent cycle. If there could be, then display the (k − 1)- and k-stems on the screen and ask the user if there are any more differentials. If the answer is yes, then go to D; otherwise go to E.

Step D. Ask the user to specify the source and target of the next differential. Record it as in 1.4. Go to C.

Step E. See if the (k − 1)-stem has any fertile elements. If it does not, then go to I; otherwise go to F.

Step F. See if there are pairs X,Y in the (k − 1)-stem such that Y is alive on S^{2pm-1} and X is alive on S^{2pm+1} for some m. If there are none, then go to H; otherwise go to G.

Step G. Look for pairs X,Y in the (k − 1)-stem such that HI(X) = HI(HI(Y)). For each such pair set PM(X) = Y. Ask the user if any values of PM should be changed and record such changes.

Step H. Set SH(X) = HI(X) for each X in the (k − 1)-stem. Ask the user if any value of SH should be changed and record such changes.

Step I. If the (k − 1)-stem is nonempty then print it on paper.

Step J. Increase k by 1 and go to A.

4. EMPIRICAL OBSERVATIONS ON THE GROWTH OF HOMOTOPY GROUPS

In this section we discuss the size of the homotopy groups of spheres. We fix a prime p throughout and consider only the p-components of the groups in question. We consider the stable case first.

Definition 4.1

Let S(k) denote the base p logarithm of the order of the direct sum of all the stable homotopy groups of spheres in positive stems $\leq$ k. Let SJ(k) denote the corresponding function for the image of the J-homomorphism, and let SV(k) = S(k) — SJ(k).

We wish to examine the growth of these functions. Since Im J is known, SJ(k) is known for all k and is easily seen to be asymptotically linear.

More precisely, if we filter Im J by powers of p, we get one generator every q stems (where q = 2p — 2 as usual), a second generator every pq stems, a third generator every p^2q stems, and so on. Therefore SJ(k) is approximately

$$k/q + k/pq + k/p^2q + \cdots = pk/q(p - 1) \qquad [4.2]$$

In [5] we computed the stable stems for p = 3 and 5 through the 106- and 1000-stems, respectively. We found that the function SV(k) is roughly quadratic, as Table 1 illustrates.

Note that the increasing value of the ratio of k^2 to SV(k) means that the growth of SV(k) is slightly less than quadratic.

Now we turn to the unstable case. For each odd-dimensional sphere S^n we could define functions S(n,k), SJ(n,k), and SV(n,k) similar to those of Definition 4.1. However, the following functions are more useful for estimating the computational difficulty of the problem we are considering.

Definition 4.3

Let U(k) denote the number of output basis elements in positive stems $\leq$ k for the program described in Section 1. Let UV(k) be a similar

Table 1 The Function SV(k) for Small Values of k for p = 3 and 5

k	SV(k)	$k^2/SV(k)$
	(a) p = 3	
20	3	133
30	7	129
40	12	133
50	19	132
60	21	171
80	29	220
106	55	204
	(b) p = 5	
50	2	1250
100	6	1670
150	12	1880
200	23	1740
250	42	1490
300	53	1700
400	78	2050
500	126	1980
600	169	2130
700	224	2190
800	284	2250
900	352	2300
1000	420	2380

function for the program as modified in Section 2, and let UJ(k) = U(k) − UV(k).

Thus U(k) is the number of permanent cycles in the EHPSS, excluding the infinite summands; UJ(k) is the corresponding number for the portion of the EHPSS described by Gray, Mahowald, and Thompson; and UV(k) is the number for the EHPSS as modified in Section 2.

Definition 4.4

W(k), WJ(k), and WV(k) are the numbers of basis elements created in the three variants of the EHPSS. Equivalently, W(k) is the base p logarithm of the direct sum of all the groups in the E_1-term of the EHPSS (excluding the infinite summands) in stem $\leq$ k.

There is a predictable relation among these functions.

Proposition 4.5

1. W(k) + S(k) = 2U(k) and similar equations hold for the other functions.
2. S(k) is approximately (p − 1) times the derivative of W(k).

Proof. Every basis element in the E_1-term is either a source of a differential, a target of same, or a stable element. If T(k) is the number of targets then we have

$$U(k) = T(k) + S(k)$$

while

$$W(k) = 2T(k) + S(k)$$

from which part 1 easily follows.

For part 2, note that most of the groups in the E_1-term are stable. For purposes of approximation, assume that they all are. Then the E_1-term in dimension n consists of the direct sum of these stable groups from two out of every q stems less than k. The dimension of this vector space is the derivative of W(k), which gives the result.

Table 2 The Function UV(k) for Small Values of k for p = 3 and 5

k	UV(k)	$k^2(k + 3)/UV(k)$	(k + 3)SV(k)/UV(k)
		(a) p = 3	
20	6	1530	11.5
30	20	1490	11.6
40	48	1430	10.8
50	89	1490	11.3
60	159	1430	8.3
		(b) p = 5	
50	3	44200	35.3
100	27	38100	22.9
150	90	38300	20.4
200	212	38300	22.0
250	421	37600	25.2

Table 3 The Functions UV(k) and UJ(k) for Small Values of k for p = 3 and 5

k	UV(k)	UJ(k)	UA(k)	U(k)
		(a) p = 3		
20	6	22	30	28
30	20	48	104	68
40	48	82	241	130
50	89	127	513	216
60	159	180	897	339
		(b) p = 5		
50	3	28		
100	27	105		
150	29	231		
200	212	406		
250	421	630		

Similar relations hold for the functions UV(k) and UJ(k). The data of Table 1 suggest that SV(k) is approximately k^2/c for some constant c depending on p. Combining this with 4.5 and some elementary calculation gives the following.

Corollary 4.6

The following approximations hold in low dimensions for odd primes.

1. $UV(k) \approx (k^3 + 3k^2)/6(p - 1)c$.
2. $SV(k)/UV(k) \approx 6(p - 1)/(k + 3)$.
3. $SV'(k)/UV'(k) \approx 4(p - 1)/(k + 2)$.

Note that part 2 shows which portion of the output basis consists of stable elements and part 3 gives the same ratio in the k-stem. Now we will tabulate the observed values of the function UV(k) in the modest range in which we have calculated (Table 2). The number in the fourth column should be compared to the value $6(p - 1)$ (i.e., 12 for $p = 3$ and 24 for $p = 5$) predicted by Corollary 4.6.

Next we compare the functions UV(k) and UJ(k) (see Table 3). The latter can be computed theoretically by combining equation (4.2) and Corollary 4.6. The resulting formula is

$$UJ(k) = pk(k + q)q^3 \qquad [4.7]$$

The function UJ is quadratic while UV is presumably cubic, so for large k UV should dominate UJ, but for our range of k UJ is larger. This comparison shows how much effort is saved by the modifications of Section 2.

For $p = 3$ we can compare U(k) with UA(k), the number of basis elements in the unstable Adams E_2-term, as computed in [7].

REFERENCES

1. E. B. Curtis, P. Goerss, M. E. Mahowald, and R. J. Milgram, Calculations of unstable Adams E_2-terms for spheres, Algebraic Topology Proceedings, Seattle 1985, Lecture Notes in Math. 1286, Springer-Verlag, New York, (1987), 208–266.

2. B. Gray, Unstable families related to the image of J, Math. Proc. Camb. Philos. Soc., 96, (1984), 95–113.

3. M. E. Mahowald, The image of J in the EHP sequence, Ann. Math. 116, (1982), 65–112.

4. M. E. Mahowald and M. C. Tangora, An infinite subalgebra of $Ext_A(Z_2,Z_2)$, Trans. Am. Math. Soc. 132, (1968), 263–274.

5. D. C. Ravenel, Complex Cobordism and Stable Homotopy Groups of Spheres, Academic Press, Orlando, Fla. (1986).

6. D. C. Ravenel, Streamlining the EHP sequence by excluding v_1-periodicity, to appear.

7. M. C. Tangora, Computing the homology of the lambda algebra, Mem. Am. Math. Soc. 337, (1985).

8. R. Thompson, The v_1-periodic homotopy of unstable spheres at odd primes, to appear.

13

A Computer Language for Topologists

DAVID L. RECTOR

University of California at Irvine
Irvine, California

This chapter discusses a computer algebra system that I am constructing for use by algebraic topologists. The current phase of the construction, the underlying computer language, is inspired by SCRATCHPAD II built by Richard Jenks and others at IBM Thomas J. Watson Research Center [1]. This will also be an advertisement of that system, which is the first symbol manipulator that promises to be of general use in topology. Unfortunately, SCRATCHPAD is not yet a product and will run only on large IBM computers. I am trying to reproduce some of its capabilities in a system that will run on small computers as a basis for my own experiments in symbol manipulation. I have taken the liberty of naming my system POINCARÉ.

1. MOTIVATION

Most of the talks in this conference have dealt with large specialized computational problems. Such problems are relatively insensitive to the computer language employed for their solution since the work the programs perform—many man-years—is large compared with the programming effort. I am more interested in the routine computations that appear in theoretical research. These computations are a severe test of a symbolic manipulation system. First, few theoretical computations last more that a few weeks when done by hand. Most take a few hours. Thus, if a computer system is to save time and effort, applications must be programmable quickly. This calls for a very high level programming language which is powerful and supports the use of existing routines in new ways. Second, theoretical calculations demand a very fine control of the algebraic simplification procedure in order to get expressions into usable forms. One does not usually know in advance what form an answer should take.

Consider the following very simple hand calculation. This is Atiyah's "postcard proof" of Hopf invariant one, which has inspired many important calculations in topology. The calculation arises from this situation. Let X be a space such that

$$H^*(X;Z) = Zx + Zy$$

where $\deg(x) = 2n$, and $\deg(y) = 4n$. Then $x^2 = ay$, where a is the Hopf invariant (up to sign). It follows that K(X) has two generators, which we shall also call x and y, and the Adams operations are given by

$$\psi^2(x) = 2^n x + ay, \qquad \psi^2(y) = 2^{2n} y,$$

$$\psi^3(x) = 3^n x + by, \qquad \psi^3(y) = 3^{2n} y.$$

Here is the calculation:

1. Compose ψ^2 and ψ^3; expand the result.

$$\psi^3\psi^2(x) = 2^n 3^n x + 2^n by + 3^{2n} ay$$

$$\psi^2\psi^3(x) = 3^n 2^n x + 3^n ay + 2^{2n} by.$$

Most computer algebra systems are happy to do this. Indeed, many systems automatically write expressions in their most expanded form. Unfortunately, expansion destroys factoring information which may be crucial theoretically and may be expensive to recover later.

2. Subtract and collect terms on x and y.

$$(\psi^3\psi^2 - \psi^2\psi^3)x = (2^n b + 3^{2n} a - 3^n a - 2^{2n} b)y$$

This is less easy; many systems will prefer to collect on a and b, depending on the order in which the symbols a, b, x, and y were encountered by the system.

3. Pick off the coefficient of y and collect terms on a and b.

$$(3^{2n} - 3^{n})a - (2^{2n} - 2^{n})b$$

4. Factor.

$$3^{n}(3^{n} - 1)a - 2^{n}(2^{n} - 1)b$$

It is hard to tell most systems how to do this.

The proof now proceeds by noting that for this expression to be zero with $a \equiv 1 \pmod 2$, 2 must divide $3^n - 1$, which can happen only if $n = 1$, 2, or 4 by a result in elementary number theory.

There are three general ways to control a simplifier to produce such calculations; all of these ways are needed. First, the desired operations may be indicated in structural terms, as in step 3: "pick off the coefficient of y." This suggests that one should have a screen-oriented expression editor which would allow a user to point to a subexpression to indicate the desired operations. Expression editors exist for LISP but not, to my knowledge, for any algebraic system.

Second, one may specify which algebraic transformations are to be applied in reducing expressions. For example, consider the rule "distribute monomials over sums," whose inverse is "factor common monomials." Application of this rule and its inverse produce important equivalences such as

$$1 + x(1 + \cdots + x(1 + x))\cdots) = 1 + x + x^2 + \cdots + x^n$$

The inverse of this rule might be used to produce step 4 of the Atiyah calculation. Unfortunately, controlling algebraic transformations is a crude tool without some way to indicate which variables or parts of expressions are to be subjected to the transformations. For example, one might wish to expand an expression on the principal variable while keeping coefficient expressions factored.

The third way to control simplification, and the way that concerns us here, is semantically. To perform Atiyah's calculation, we want to write the expression as an element of (say) a free abelian group with generators x and y with coefficients in the ring of polynomials in a and b with coefficients in integer expressions. If we can specify this semantic structure for our calculation, it is then easy to apply other methods of control selectively: "factor all monomials from integer expressions."

To construct simplifiers that are controlled semantically, we must have an underlying computer language that understands mathematical structures. This turns out to be related to another important consideration. Much of the calculation of topology is carried out over peculiar coefficient rings—such as integers mod p, factored integers, and finite fields. Most existing computer algebra systems were designed with physicists in mind and only recognize subrings of the real numbers. When we go to the trouble to write a program to manipulate, say, polynomials, we want that program to work for all suitable coefficient rings. Programs of that kind are called generic routines.

2. REQUIREMENTS OF A LANGUAGE

The minimal language features required for symbol manipulating in pure mathematics are

1. Parametrized data types
2. Overloading of operators
3. Generic routines
4. Pointer-oriented data structures with automatic garbage collection
5. Interactive capability

The last two are characteristic of all existing general-purpose computer algebra systems and their underlying LISP or LISP-like programming languages. Data structures must consist of pointers since a mathematical expression is naturally represented as a tree structure and cannot be assigned storage in advance of its generation. Interactive capability is implied by the uses to which these systems are put and is also important to the rapid development of applications.

The first three features are new; the only computer algebra system incorporating them is SCRATCHPAD II. Let us illustrate these notions by an example. Suppose we wish to write programs to manipulate polynomials. We want to construct a new data type or algebraic domain—that is, a set of data together with some operations on the data—which we will denote by Poly. A mathematician immediately expects to specify a ring R of coefficients. Thus the data type Poly should come with a parameter R indicating the type of the coefficients: Poly[R]. Since the system should do as much type checking as possible, the algebraic properties—operations and axioms—of the coeffficient type R should be specified in some way. To abbreviate that specification, Richard Jenks [1] has introduced the important notion of the category of a data type, which consists of a list of operations and a list of axioms. Thus when we define Poly we will declare that R is to be a Ring: Poly[R : Ring].

Now, the operations of all rings recognized by the system must be denoted by the same names so that the rings are interchangeable (overloading of operators). For the convenience of the user, the ring operations should use the conventional names +, *, etc. To distinguish the operations of Poly[R] from those of R, when this is not clear from context, we need a notation: Poly[R]$+ and R$+. When R is given—for example, if one envokes the data type Poly[Integer]—the system must fill in the necessary references to the operations of R—Integer$+, Integer$*, etc.—in the programs of Poly.

Here is an abbreviated specification of the type generator Poly. The language used here is approximately that which I expect in my finished system except that keywords are given in boldface for readability.

```
Poly == cluster[R: Ring]
        is Ring
          with
            create: list(R) -> $
            coef: $ -> list(R),
            ...,
            if R has gcd: (R, R) -> R then
              gcd: ($, $) -> $
              end
        rep == array[R]
        plus == proc(a: cvt, b: cvt)returns(cvt)
                i,na,nb,m,n: Integer
                na := rep$high(a)
                nb := rep$high(b)
                m := Integer$min(na, nb)
                n := Integer$max(na, nb)
                c: rep := rep$predict(0, n)
                for i in [0..m] do
                  rep$addh(c, a[i] R$+ b[i])
                  end
                if nb > na then a := b end
                for i in [m + 1 .. n] do
                  rep$addh(c, a[i])
                  end
                return(c)
                end plus
        ...
        end Poly
```

Notice a few things. Following the is is a description of the category of the type to be constructed listing all the operations of the type. A category is analogous to the mathematical notion of a category. The type generator shown here is analogous to a functor. The phrase "if R has gcd..." allows the type to be generated by Poly to vary according to the characteristics of R. The symbol "$" appearing in this description is shorthand for the type to be generated. The description of the parameters and the target category is the public part of the type generator. It is only part of the specification that is available to outside programs.

Following the public part is the private part of the specification, which begins with the representation of the data. The representation is not available outside Poly. It is very important in the construction and maintenance of programs that one part of a program not be able to meddle in the affairs of another. For example, some of the procedures in Poly would crash if passed an empty array of coefficients. This cannot happen, however, for a polynomial can only be constructed using the routines of Poly. No other part of the system can modify the array of coefficients. In addition, the data representation can be changed—for example, to improve efficiency—without changing routines outside Poly.

Finally, the code for the various operations is given. The code for plus, to be substituted for Poly$+, is shown here. The **cvt** type declaration for the variables of plus indicates that these variables are to be viewed in the internal representation (as arrays of elements of R). Only a few routines in Poly will need to use the internal representation.

3. CLU

The language that is to form the basis of my system is an extension of CLU. CLU was developed as a model high-level language by Barbara Liskov and others at MIT [3]. Its major design consideration was to permit user-defined data types to be used as easily as system-defined ones. CLU is not interactive, so I am adding features borrowed from SCRATCHPAD—notably, categories—that are necessary for interactive use. I am also making some notational changes to bring CLU into line with mathematical usages, but I am trying to preserve the simplicity of the original language. I can highly recommend CLU, even as it stands, for mathematical computation. It is available on VAX UNIX systems as well as some smaller workstations.

CLU is a pointer-oriented language like LISP. All variables in the system are pointers to data objects. When no variable points to a data object, it ceases to exist and is garbage-collected. A CLU program consists of

1. Procedures
2. Iterators
3. Clusters (data type generators)

Everyone is familiar with procedures. The major innovation is that a CLU procedure may have parameters so that the parametrized types in it may be assigned values. The iterator, on the other hand, is a powerful innovation which allows the traditional **for** statement to be used with any user-defined data type. An iterator is a procedure which, once entered, yields a sequence of values. As an example, consider the following iterator, which yields all the values on a binary tree:

```
nodes == iter[n: type](T: tree[n])yields(n)
         where tree[n] has
           left: proctype(tree[n])returns(tree[n])
           right: proctype(tree[n])returns(tree[n])
           empty: proctype(tree[n])returns(bool)
           value: proctype(tree[n])returns(n)
       tr == tree[n]
       if tr$empty(T) then
          yield(tr$value(T))
          for v: n in nodes[n](tr$left(T)) do
             yield(v)
             end
          for v: n in nodes[n](tr$right(T)) do
             yield(v)
             end
          end
       end nodes
```

Suppose now one has a tree T of type Name, such as

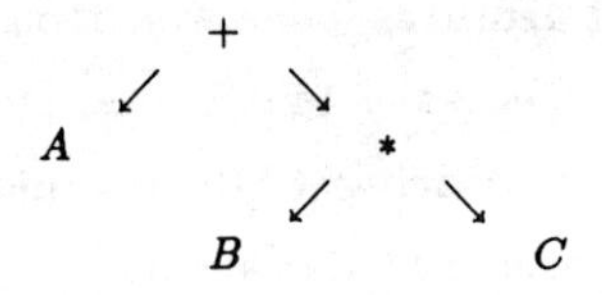

which is a representation of the algebraic expression A + B * C, and one wishes to print out the nodes of T: +, A, *, B, C. One would write

```
for v: Name in nodes[Name](T) do
   IO$put_name(v)
   end
```

The tree T will be passed as a variable to nodes. The statement yield(tr$value(T)) causes the value of T (the top node) to be computed. At this point, the execution of nodes is suspended, the value yielded is assigned to v, and the statement after the do is executed—i.e., v is output. When the execution of the do is complete, nodes is resumed to yield the rest of its values: the nodes of the right branch and the nodes of the left branch of T. Notice that nodes calls itself recursively.

4. SYSTEM DESIGN

There are two interesting technical problems in building an interactive computer algebra system that I would like to discuss briefly. Both of these have been solved in the SCRATCHPAD II system, but the algorithms are not published. First, when writing code to be compiled, one wants one's language to be strongly typed; all type conversions should be explicitly written out. The reason for this is error detection. By the time one gets a strongly typed language to compile, most of one's errors have disappeared. On the other hand, one wants an interactive language to be more forgiving. In constructing SCRATCHPAD II, the IBM group has tried to make the system do as much educated guessing about the user's intentions as possible in interactive use. This leads to two languages: a "conversational" language with lots of automatic type conversions and a strongly typed algorithmic language for archived programs. I intend to require more direction from the user than SCRATCHPAD. First, I must keep the programming effort within bounds. Second, algebraic topology uses a large variety of mathematical constructs, while SCRATCHPAD is oriented closely to ring theory. I need a simple system of type assumptions that users can augment easily to serve their own needs.

Type analysis of expressions works like this: consider the expression

$$(1 + z) * (x + I * y)$$

The outermost operation is * with two operands, (1 + z) and (x + I * y). First, type the operands by recursion; suppose them to have types T_1 and T_2. There may be no * operation between T_1 and T_2. Thus one must find domains T_1' and T_2' such that there are sequences of type conversion maps coerce : $T_1 \to T_1'$ and $T_2 \to T_2'$ and also an operation $T_1' * T_2' \to T_3$ for some domain T_3. The T_1', T_2', and T_3 must be chosen to be minimal in some lattice of domains. When one comes to the subexpressions (1 + z) and (x + I * y), the possibilities for ambiguity become apparent. Is z to be considered a polynomial in one variable, a multivariant polynomial, or a general expression? Is z to be an integer, integer mod p, real or complex number? These ambiguities must be resolved by a well-conceived set of defaults or by user direction.

Constructing the data base for automatic typing is interesting in a language with parametrized types, for parametrized data type generators tend to produce an infinite data base. For example, suppose we have rings and diagrams of coerce maps

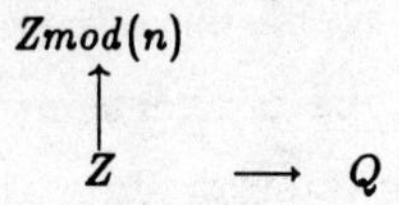

$$\begin{array}{ccc} Zmod(n) & & \\ \uparrow & & \\ Z & \longrightarrow & Q \end{array}$$

where Zmod(n) is the congruence ring mod n, and Z and Q are the integer and rational numbers. Then the type generator Poly produces the further diagrams

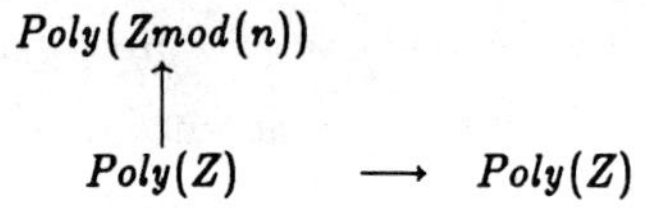

$$\begin{array}{ccc} Poly(Zmod(n)) & & \\ \uparrow & & \\ Poly(Z) & \longrightarrow & Poly(Z) \end{array}$$

of rings, and so on. There is a further complication from the maps coerce : R → Poly(R) which are provided by the type generator Poly. One way to solve this problem is to encode the data base as a simple Horn clause logic in which the type generators are functions and the operators are predicates. Choices of type conversions can then be made using standard automatic theorem-proving algorithms.

The second problem, which is related to the first, is that of efficiency in the type analysis. The first attempt by IBM to put strong typing into an algebraic language was MODLISP [2]. In MODLISP each list was assigned a type and the MODLISP interpreter analyzed types for each operation. MODLISP was inefficient. The difficulty was that when a loop (or equivalent recursion) was executed, the type analysis was done on every trip through the loop. Thus, as far as possible, the type analysis of expressions must be done statically before any evaluation is attempted. This amounts to a partial compilation of expressions.†

My system is organized as follows:

1. A parser translates input expressions into a simple tree structure of operator followed by operand list (syntactic analysis).

---

†I am indebted to R. Jenks for this information.

2. The tree is typed (semantic analysis) with either strong type checking or automatic type conversions. The result is a "decorated tree."
3. The decorated tree is encoded. In the interactive mode, the result is a tree that may be directly interpreted. In compiled mode, the result is a LISP or native code program.

I don't know how efficient all this is going to be until I try it.

I am implementing my system in Cambridge LISP on a Motorola 68000-based computer system with 1 MB of main memory. The underlying operating system is a single-user multitasking system, which is very convenient for program development since one may edit source code while keeping an interactive test environment intact. The Cambridge LISP system is available on the AMIGA computer. That machine should be available with sufficient memory (at least 1 MB) to run my system by the time it is ready.

Since my system is being developed for testing language ideas, I will be using a compiler generator to construct the parser. A compiler generator takes an abstract description of the language grammar (Backus-Naur form) and produces a program (in LISP in this case) to parse it. Thus extensive language changes can be accomplished in a few hours. I have almost completed the coding (in LISP) of the compiler generator.

REFERENCES

1. R. D. Jenks, A primar: 11 keys to New Scratchpad, EUROSAM 84, Lecture Notes in Computer Science, 174, Springer-Verlag, New York (1984), 123–147.
2. R. D. Jenks, MODLISP: an introduction, Symbolic and Algebraic Computation, Lecture Notes in Computer Science, 72, Springer-Verlag, New York (1979), 466–480.
3. Barbara Liskov et al., CLU Reference Manual, Lecture Notes in Computer Science, 114, Springer-Verlag, New York (1981).

14

Parabolic Representations and Symmetries of the Knot 9_{32}

ROBERT F. RILEY

State University of New York at Binghamton
Binghamton, New York

We outline a proof that that knot k of type 9_{32}, drawn in Figure 1a, is asymmetric; i.e., every autohomeomorphism of its complement $S^3 - k$ is isotopic to the identity. As in [4], we do this by exhibiting a fundamental domain $\mathfrak{D} \subset \mathfrak{U}^3$, the upper half-space model of H^3, for the image of a faithful discrete representation of the knot group, which admits no hyperbolic symmetries. Our knot 9_{32} is not the first known example of an asymmetric knot. In 1981 W. Thurston drew the diagram Figure 1b and labeled it a "totally asymmetric knot."

Currently the only serious obstacle to exhibiting the hyperbolic structure for a randomly selected excellent (hyperbolic) knot by this author's method is the difficulty of determining the equivalence classes of parabolic representations (p-reps) of its knot group. For 2-bridge knots we showed in [1] that these equivalence classes correspond to the irreducible factors of a certain easily computed integral polynomial. We have no similar result for most knots of higher bridge number, and instead we have to express equivalence classes as solutions to systems of nonlinear integral polynomial equations on several variables. In [4] we outlined how we do this using the SAC-1 computer algebra system, but for the past 4 years this has not been available to us. At the "Computers in Geometry and Topology" Conference at the University of Illinois in Chicago in March 1986 we learned that computational commutative algebra is now an actively evolving subject, so we can hope that solving

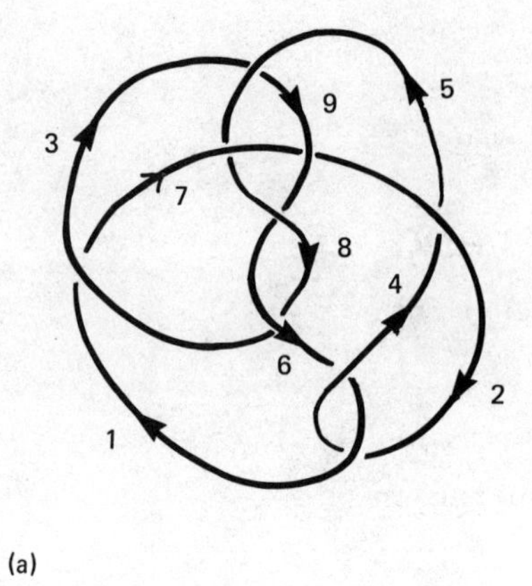

(a)

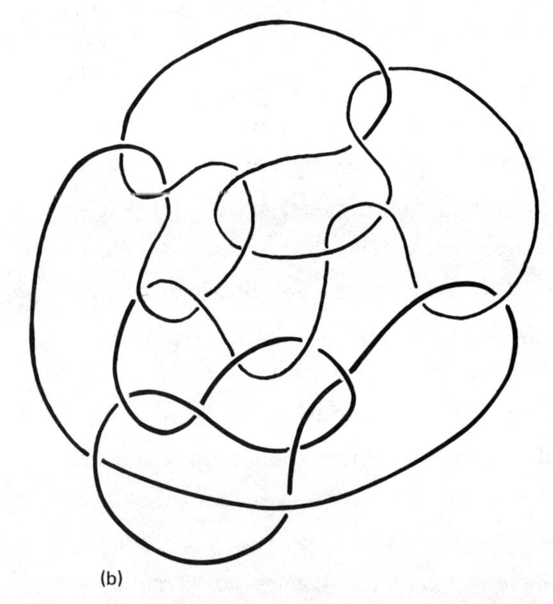

(b)

Figure 1 (a) 9_{32}; (b) Thurston's knot.

systems of polynomial equations will become much easier, if not routine. Right now, however, solving a polynomial system is such an adventure in each case that it would be a foolish project to determine all equivalence classes of p-reps for all prime knots of 9 or 10 crossings. (The results for ≤ 8 crossings are now complete.) One curse on the theory of hyperbolic structures for knot complements is that the precise algebraic description of the hyperbolic structure is likely to involve algebraic number fields of very high degree and rational numbers with huge numerators and denominators; see the summary of the results for 9_{32} in Section 2 below. But once one has the p-reps for a knot group, deciding which p-rep might give the hyperbolic structure and then computing a fundamental domain etc. for it has been an easy routine for several years; see [4] or the results for 9_{32} in Section 3 below.

The specific instigation for this note came from meeting David Bayer and Michael Stillman at the Chicago conference and getting a copy of their Macaulay computer algebra system for a Macintosh home computer. This substitutes for and in some ways improves on the SAC-1 system. In Section 1 we explain how to use it to do what once had been the hardest part of equation solving. But this is only the initial step, and the other steps were carried out on the author's home computer, a Hewlett-Packard HP-71B, the SUNY-Binghamton's main IBM computer, and finally a Microvax II computer running Macsyma in our mathematics department that was partly funded by an NSF grant. I wish to thank my colleague Dr. Frederick Sullivan for the prolonged hard work he devoted to making this computer available and for his help with Macsyma.

1. SETTING UP AND SOLVING SYSTEMS OF POLYNOMIAL EQUATIONS

In [4, section 5] we gave a brief account of how to find the equivalence classes of p-reps for a 3-bridge knot group πK. Here we give a modified account that describes how to use Macaulay in place of SAC-1 and gives more detail on certain parts of the calculation.

First of all we choose an over presentation for πK with three meridian generators m_1, m_2, m_3 and two relations:

$$\pi K = |m_1, m_2, m_3 : r_1(\vec{m}), r_2(\vec{m})| \qquad [1.1]$$

Here $\vec{m}$ abbreviates (m_1, m_2, m_3), and the relations r_1, r_2 will be expressed in one of two ways. In the first way r_ν is $w_\nu m_a = m_b w_\nu$, where $m_a, m_b \in \vec{m}$ and w_ν is a certain word on $\vec{m}$ (this is a 3-bridge presentation). The second way uses a sequence of meridians $m_4, \ldots,$ m_n of πK defined by assistant relations, and then the relation r_ν is written $m_a m_b = m_c m_a$ for certain meridians. The first form of group presentation is obtained from the second by a straightforward elimination-of-bridges process, but one needs a computer and a tested computer program to do this reliably.

To find the equivalence classes of p-reps θ of πK such that $m_1\theta \not\rightleftarrows m_2\theta$ (does not commute with) we substitute $\vec{P}$ for $\vec{m}$, where

$$P_1 = \begin{bmatrix} 1 & 1 \\ 0 & 1 \end{bmatrix}, \quad P_2 = \begin{bmatrix} 1 & 0 \\ -x & 1 \end{bmatrix}, \quad P_3 = \begin{bmatrix} 1 - yz & y^2 z \\ -z & 1 + yz \end{bmatrix} \qquad [1.2]$$

in the relations r_1, r_2 to get two matrix equations and hence eight entry equations on x, y, z with integral coefficients. But actually only two entries are needed per relation, because the relations assert that conjugate things are equal, and the normal form (1.2) already ensures that each meridian image has trace two and determinant one. So we obtain a system of polynomials $p_\mu(x,y,z) \in Z[x,y,z]$, $\mu = 1,\ldots,4$, the <u>p-rep polynomials</u> for the presentation (1.1) and normal form (1.2). The <u>p-rep equations</u> are $p_\mu(x,y,z) = 0$ for all μ.

To solve the p-rep equations using SAC-1 one would eliminate variables by taking resultants, as discussed in [4, section 5]. The Macsyma manual describes a function, algsys, which purportedly works the same way. But beware that resultants tend to have extraneous factors, and that after eliminating one variable all the resultants eliminating the next variable might vanish. When this happens one must calculate greatest common divisors of these multivariate polynomials and explore a tree of possible alternatives, an enormous complication of this method. Using Macaulay, one eliminates variables by computing a standard (Gröbner) basis for the ideal generated by the p-rep polynomials and separating out the part of the basis that doesn't involve the variables to be eliminated. This process seems to work more reliably, but because Macaulay is not a complete algebra system there are certain complications. One of these is that a standard basis can be computed only for an ideal of homogeneous polynomials. To get homogeneous p-rep equations we replace $\vec{P}$ of (1.2) by $\vec{Q}$, where

$$Q_1 = \begin{bmatrix} 1 & 1 \\ 0 & 1 \end{bmatrix}, \quad Q_2 = \begin{bmatrix} t & 0 \\ -x & t \end{bmatrix}, \quad Q_3 = \begin{bmatrix} t^3 - tyz & y^2z \\ -t^2z & t^3 + tyz \end{bmatrix} \qquad [1.3]$$

Because Q_2, Q_3 are not invertible over Z[t,x,y,z] (and because Macaulay cannot invert an invertible matrix) we substitute quasi-inverses

$$Q_1^* = \begin{bmatrix} 1 & -1 \\ 0 & 1 \end{bmatrix}, \quad Q_2^* = \begin{bmatrix} t & 0 \\ x & t \end{bmatrix}, \quad Q_3^* = \begin{bmatrix} t^3 + tyz & -y^2z \\ t^2z & t^3 - tyz \end{bmatrix}$$

for inverses, where for each Q, $QQ^* = Q^*Q$ is a scalar matrix. Each time a matrix Q is defined or calculated one must similarly provide Q^* if it will be needed. Thus one can set up homogeneous p-rep polynomials $p_\mu(t,x,y,z)$, $\mu = 1,\ldots,4$, and the corresponding equations. By ordering the variables in three different maps before eliminating (using a ring homomorphism facility in Macaulay), we eliminate two variables, and then set t = 1 (by another ring homomorphism) to obtain one-variable p-rep equations

$$F_1(x) = 0, \qquad G(y) = 0, \qquad H(z) = 0 \qquad [1.4]$$

where F_1, G, H have integral coefficients. But there is a second complication here: Macaulay does all arithmetic modulo a prime π, so the p-rep polynomials p_μ actually belong to $F_\pi[t,x,y,z]$ where $F_\pi = Z/(\pi)$, and the polynomials (1.4) are reduced modulo π. Furthermore, Macaulay makes all polynomials monic for the characteristic π it is using. Fortunately, π can be chosen, and the largest allowed values are

$$31991, \qquad 31981, \qquad 31973 \qquad [1.5]$$

If one has reductions modulo several primes of an integral polynomial p one may reconstruct p using the Chinese Remainder Theorem. When p has rational coefficients it seems necessary to guess the denominators and multiply by them before Chinese remaindering. Therefore if one of F_1, G, H is not monic, one must first guess the highest coefficient using whatever supplementary information is at hand. So far, the troubles caused by this complication are small in comparison to the advantage of avoiding resultants.

The primes (1.5) are not nearly as large as we would like, but they are much larger than the numerical coefficients of any system of p-rep equations encountered up to now. This allows us to set up the p-rep equations in characteristic π_1 and later convert them to characteristic π_2 by a ring homomorphism without getting nonsense. We note also that an integral polynomial purportedly reconstructed from its reductions modulo several primes is not guaranteed to be the correct reconstruction unless one has an a priori bound on the magnitude of its coefficients. We have no such bound here, but it is not necessary to prove that our polynomials (1.4) are correct when they actually are but we don't know it. The final step of this long process will be the formal verification by direct substitution of the presumed solution in the p-rep equations that the answer is correct, and this makes the separate verification of the several steps of the process redundant.

Suppose that we have obtained correct polynomials (1.4) and that these are irreducible of the same degree, say d. (This is the case for 9_{32}, and if the degrees do differ or the polynomials are reducible the modifications to this discussion are straightforward.) Each p-rep θ of πK captured by (1.2) corresponds to a triple of values $x = x(\theta)$, $y = y(\theta)$, $z = z(\theta)$ of the variables. The field of θ, $F(\theta)$, is $Q(x(\theta), y(\theta), z(\theta))$, and it has d imbeddings in C corresponding to d triples (x_1, y_1, z_1) of complex roots of the polynomials (1.4). These triples can be determined numerically by simply substituting all the possibilities in one of the p-rep polynomials and keeping the triples that make it approximately vanish.

Our hypotheses imply that $F(\theta) = Q(x(\theta))$, so that $y(\theta)$, $z(\theta)$ can be expressed as rational polynomials of formal degree $d - 1$ on $x(\theta)$, say by

$$y = F_2(x), \qquad z = F_3(x), \qquad F_2, F_3 \in Q[x] \tag{1.6}$$

Then the complete algebraic description of this equivalence class of p-reps is the triple (F_1, F_2, F_3), which has the property that the substitution of $F_2(x)$ for y, $F_3(x)$ for z in any p-rep polynomial produces a polynomial in Q[x] which is divisible by $F_1(x)$.

One can determine F_2, F_3 using approximate arithmetic as follows. If (say) $y = F_2(x)$, then for each root triple (x_1, y_1, z_1) we have $y_1 = F_2(x_1)$. This means that the unknown coefficients $c_0, c_1, \ldots, c_{d-1}$ of F_2 are the solution of the system of d linear algebraic equations on d unknowns

$$\begin{bmatrix} 1 & x_1 & x_1^2 & \cdots & x_1^{d-1} \\ & & \cdots & & \\ 1 & x_d & x_d^2 & \cdots & x_d^{d-1} \end{bmatrix} \begin{bmatrix} c_0 \\ c_1 \\ \vdots \\ c_{d-1} \end{bmatrix} = \begin{bmatrix} y_1 \\ y_2 \\ \vdots \\ y_d \end{bmatrix}$$

Given approximate numerical values for the x_l, y_l we derive approximate numerical values c_j^* for the rational coefficients c_j. We may try to determine the exact fraction c_j by expanding the real part of c_j^* as a regular continued fraction and stopping when we appear to have the correct result. Alternatively, if F_1 is monic (the case for 9_{32}), the denominator of c_j is some divisor of the discriminant $\Delta(F_1)$ of $F_1(x)$ times the highest coefficient of G(y). The discriminants of F_1, G, H of (1.4) are likely to be much too large to factor directly, but one can compute the pairwise greatest common divisors easily using Macsyma. The factor of $\Delta(F_1)$ to look at is $\Delta(F_1) \div \gcd(\Delta(F_1), \Delta(G))$, or its square root if it is a perfect square. Given the common denominator D_{xy} for the coefficients of F_2, we determine the numerators of the coefficients by multiplying $\vec{c}^*$ by D_{xy} and rounding to the nearest integers. If our computed vector $\vec{c}^*$ is accurate enough, both methods will produce F_2 correctly, but using discriminants to predict the denominator is less demanding on the accuracy of $\vec{c}^*$.

2. THE ALGEBRA FOR 9_{32}

We use the presentation for the group πK of $k = 9_{32}$ deduced from Figure 1a. This is, explicitly,

$$\pi K = |m_1, m_2, m_3 : m_4 = m_1^{-1}m_2m_1,\ m_5 = m_2m_4m_2^{-1},\ m_6 = m_4m_1m_4^{-1},$$
$$m_7 = m_3^{-1}m_1m_3,\ m_8 = m_6m_3m_6^{-1},\ m_9 = m_5m_3m_5^{-1},$$
$$r_1 : m_7m_8 = m_5m_7,\ r_2 : m_8m_6 = m_9m_8|$$

The oriented longitude l_1 such that $\langle m_1, l_1\rangle$ is a peripheral subgroup of πK is

$$l_1 = m_3m_9^{-1}m_1m_2^{-1}m_7m_6m_5^{-1}m_8m_4m_1^{-1}$$

We began by sitting before a Macintosh computer fitted with 512K of RAM for several hours and setting up the homogenous p-rep equations, using rather a lot of intricate finger work. We eliminated variables, two at a time, to obtain the one-variable polynomials (1.4). The elimination was repeated three times per polynomial using the prime moduli (1.5), and each elimination required some 12 minutes of computer time plus several minutes of ancillary work. The polynomials (1.4) for 9_{32} have degree 29, and we determined their coefficients at home later, using the peculiar language Forth to do the Chinese Remainder calculations on our HP-71B. We found, in brief,

$$F_1(x) = -1 + \cdots + x^{29}$$
$$G(y) = 839 + \cdots + 19y^{29}$$
$$H(z) = -361 + \cdots + z^{29}$$

The undisplayed coefficients have up to nine figures but have magnitude less than one-half the product of two primes of (1.5). Because we found that each coefficient determined modulo the product of two of these primes has the right value modulo the third, we were very confident our reconstructed polynomials are correct. We guessed the coefficient 19 of y^{29} by guessing that the entries of P_3 in (1.2), in particular $y(\theta)^2z(\theta)$, are algebraic integers, even though $y(\theta)$ itself is not integral.

The next step, obtaining the 29 root triples (x_1, y_1, z_1), was also carried out on the HP-71B. It should not pass unmentioned that 15 years earlier calculating the roots of such polynomials on a university main computer running standard root-finding software would have been a very uncertain enterprise. The sorting of the roots of the three polynomials was done in the obvious way. We remark that if we had had a two-variable polynomial, say $J(x,y) \in Q[x,y]$, in the p-rep polynomial ideal the sorting could have been done much faster by using J to get pairs (x_1,y_1) before going on to triples. The trouble is that the standard basis which Macaulay produced had lots of such polynomials, but we had no idea how to guess the denominators of their coefficients quickly enough for a net saving of time and effort.

The next step was the determination of the polynomials $F_2(x)$, $F_3(x)$ of (1.6). We first attempted this on an IBM 4381 using quadruple precision complex floating-point arithmetic that is equivalent to about 33 decimal figures. The root triples were recomputed to maximal accuracy using Newton's method, and we solved for the coefficients of F_2, F_3 by row reduction of the Vandermonde linear equations and got approximate vectors $\vec{c}^*$ which are good to about 26 or 27 digits. This must be nearly optimal for this quad precision, because the accuracy of the polynomial roots is reduced when evaluation near a root leads to large intermediate values that don't quite cancel out. The continued fraction method for reconstructing the exact rational coefficients failed completely for lack of sufficient accuracy, and for a while no progress could be made.

When Macsyma became available we were able to complete the determination of F_2 and F_3. First we used the discriminant method to compute the denominators D_{xy} of F_2 and D_{xz} of F_3 and got

$$D_{xz} = 45204312944516612830837, \qquad D_{xy} = 133 \cdot D_{xz}$$

This new information combined with the IBM values of $\vec{c}^*$ showed that we were looking for numerators with up to 31 digits, so that the "bigfloat" arithmetic of Macsyma using 50 decimal figures would be appropriate. But making this work turned out to be a devil of a problem with which to learn to use Macsyma, partly because of its size and partly because of the slowness of complex bigfloat arithmetic. For example, after we found ways of pumping up the accuracy of the polynomial roots and of solving the Vandermonde linear equations, these computations required some 15 hours and between 6 and 9 hours of CPU time, respectively. On multiplying the solution vectors $\vec{c}^*$ by the above denominators, each entry was within 10^{-10} of some rational integer, whence the numerators were unambiguously determined. Someone wishing to see $F_1(x)$, $F_2(x)$, $F_3(x)$ should contact me to arrange some form of electronic transmission of the data.

Verification that our solution is correct is not straightforward either, because direct substitution in a p-rep polynomial requires too much machine memory. However, Macsyma supports arithmetic in $F_\pi[x]$, where F_π is the Galois field with π elements, π prime, and the verification by substitution succeeds modulo every prime we have tried. If we can put a bound B on the magnitude of any coefficient (in Z) that could arise during the evaluation of

$$D_{xy}^m D_{xz}^n p_\nu(x, F_2(x), F_3(x)) \pmod{F_1(x)}, \qquad m = \deg_y p_\nu, \; n = \deg_z p_\nu$$

where $p_\nu(x,y,z)$ is a p-rep polynomial, then by verifying the substitution modulo primes π_μ whose product exceeds 2B we deduce the verification in $Q[x]$. On working out the details we find that $B = 2.5 \cdot 10^{543}$ suffices for all the p-rep polynomials. At this writing we have used an eclectic set of primes whose product is less than $4 \cdot 10^{56}$, and we doubt we shall continue this much further. There are fewer than 210 primes π such that verification modulo π would give a false positive result, and such a prime would have to divide all of the 116 coefficients of the four remainders after division by $F_1(x)$.

We conclude this section with two tiny results that are by-products of complications of the algebraic calculations. The top coefficient 19 of $G(y)$ means that in some algebraic sense $G(y)$ has an infinite root y_∞ in characteristic 19. But the constant term -19^2 of $H(z)$ means that in characteristic 19 H has a zero root z_0 such that $y_\infty z_0 = 0$ but $y_\infty^2 z_0$ is finite and nonzero. On working out the details of this on the HP-71B we obtain a p-rep θ_1 in characteristic 19,

$$m_1\theta_1 = \begin{bmatrix} 1 & 1 \\ 0 & 1 \end{bmatrix}, \quad m_2\theta_1 = \begin{bmatrix} 1 & 0 \\ -9 & 1 \end{bmatrix}, \quad m_3\theta_1 = \begin{bmatrix} 1 & -2 \\ 0 & 1 \end{bmatrix},$$

$$l_1\theta_1 = \begin{bmatrix} -1 & 7 \\ 0 & -1 \end{bmatrix}$$

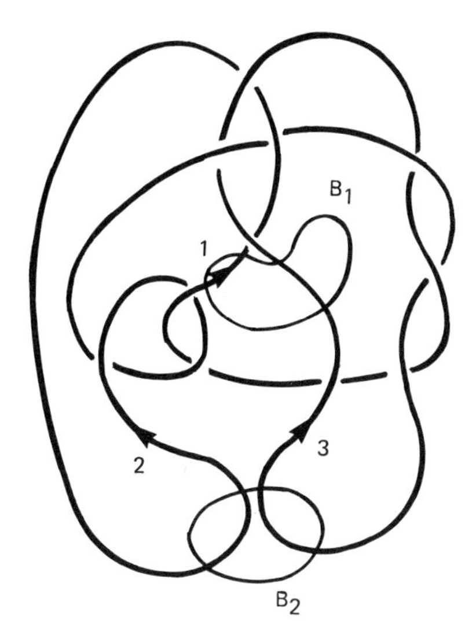

Figure 2 Another model of 9_{32}.

The constant term 839 of G(y) leads to a p-rep θ_2 in characteristic 839 with

$$m_1\theta_2 = \begin{bmatrix} 1 & 1 \\ 0 & 1 \end{bmatrix}, \quad m_2\theta_2 = \begin{bmatrix} 1 & 0 \\ 249 & 1 \end{bmatrix}, \quad m_3\theta_2 = \begin{bmatrix} 1 & 0 \\ 307 & 1 \end{bmatrix},$$

$$l_1\theta_2 = \begin{bmatrix} -1 & 62 \\ 0 & -1 \end{bmatrix}$$

The noteworthy feature of these p-reps is that $m_1\theta_1 \rightleftarrows m_3\theta_1$ and $m_2\theta_2 \rightleftarrows m_3\theta_2$. To see the implications of this, consider the model k of 9_{32} drawn in Figure 2. This indicates two 3-cells B_1, B_2 such that k meets each cell in two simple arcs. Let $k \cap B_1$ be $\alpha_1 \cup \alpha_3$, and note that these arcs correspond to the meridians m_1, m_3 of our fixed presentation of πK. Recall from [2] the notion of tangling k in B_1: we form a knot k' which coincides with k outside B_1 and $k' \cap B_1$ consists of two tame arcs with the same endpoints as the originals. The commuting trick of [2] is the remark that because $m_1\theta_1 \rightleftarrows m_3\theta_1$, θ_1 forgets how k wends through B_1, hence θ_1 induces a p-rep of the group $\pi K'$ of k'. Consequently, every knot k' obtained by this construction is nontrivial. In like manner every knot formed by tangling k in B_2 is nontrivial.

3. HYPERBOLIC STRUCTURE AND SYMMETRIES FOR 9_{32}

It is easy to obtain the hyperbolic structure once the root triples (x_1, y_1,z_1) for the polynomials (1.4) have been listed, whether or not the rest of the algebra is complete. In the present case there was really only one triple that looked at all likely to correspond to the excellent p-rep, namely

$$x_{12} = -0.720227543564297 + 1.565134811587276i$$

$$y_{12} = -0.432063252868998 + 0.899492108362272i$$

$$z_{12} = 0.806958727615393 + 2.970702809022438i$$

This and the presentation of πK displayed in Section 2 were given as data to our standard FORTRAN program, Knots3, which runs on the IBM. The results were the declaration that $\pi K\theta_{12}$ is discrete, that the side pairing transformations of a Ford fundamental domain $\mathfrak{D}$ listed in Table 1 generate it, that the edge cycle relations on these generators present $\pi K\theta_{12}$, and finally a plot, Figure 3, of the projection of $\mathfrak{D}$ on the complex plane. We regret that we had to mutilate this diagram to make it fit the page, but some of the regions in the picture are already so small that they could not be further reduced in size if they are to be seen after imperfect reproduction of the diagram. The origin of this problem is the large imaginary part of the longitude entry for this p-rep, which makes $\mathfrak{D}$ so narrow. (The longitude entry is 5.017 − 8.122i, which is changed to −0.017 + 8.122i for use as a side pairing transformation.)

Another verification which is still lacking is the check that the relations of Table 2 are consequences of the two relations of our presentation of πK. We do have a subroutine which simplifies the edge cycle relations that should help with this. But this subroutine was not very cleverly written, and in this case our simplified presentation uses non-meridian generators and is still pretty complicated.

Consider the possible symmetries of 9_{32}. As noted in [4], these correspond to EH-isometries of $\mathfrak{U}^3$ that normalize $\pi K\theta_{12}$, and there are three kinds of EH-isometries. The first is of glide-reflections which

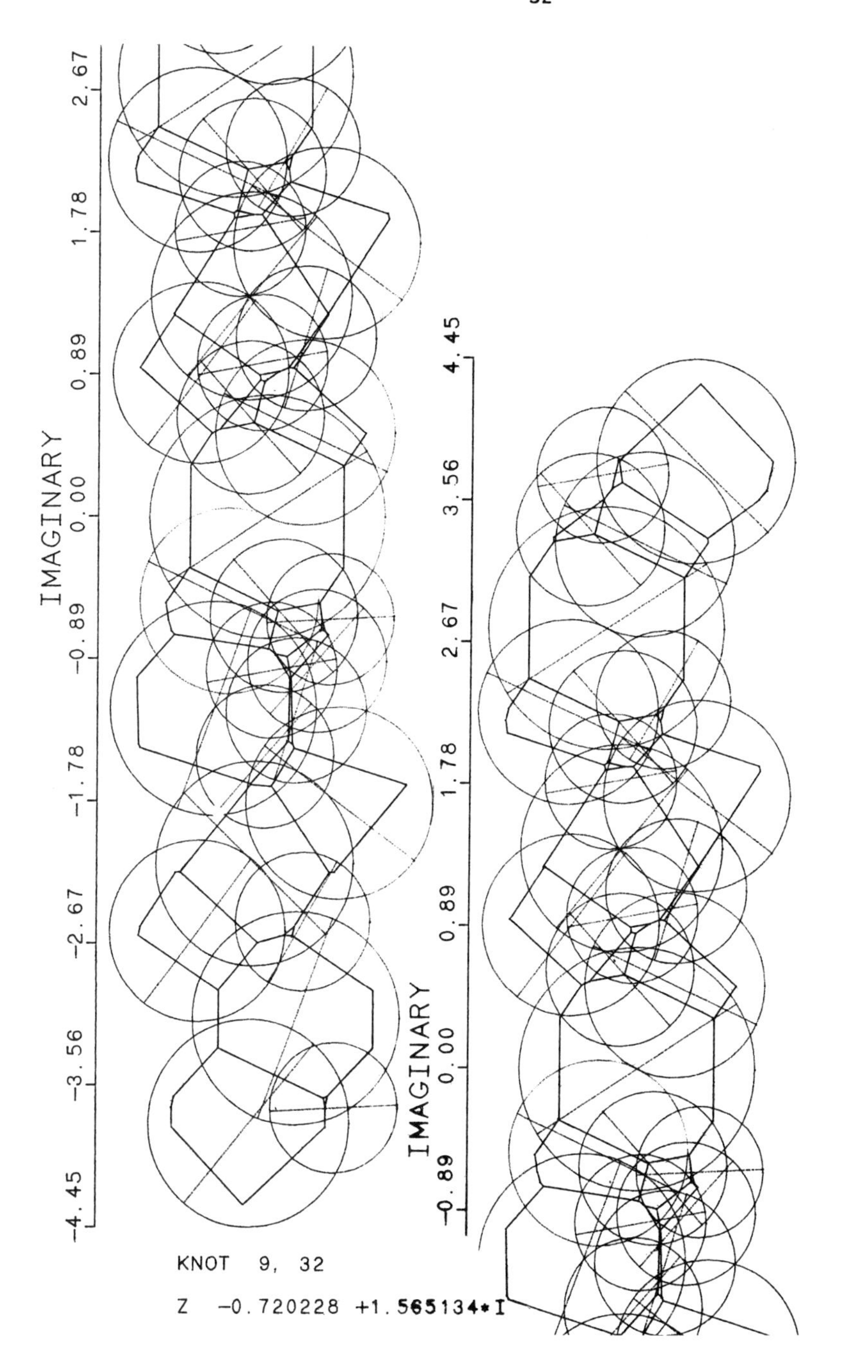

Figure 3 The projection of the fundamental domain $\mathfrak{D}$ for $\pi K\theta$ on the complex plane.

Table 1 For Each Side Pairing Transformation T_n of $\mathfrak{D}$ We List n, $rd(T_n)$, $cn(T_n)$, and $cn(T_n^{-1})$

Sphere	Radius	$cn(T_n)$	$cn(T_n^{-1})$
1	0.580	−0.243 − 0.527i	0.243 + 0.527i
3	0.762	0	−0.217 + 2.763i
5	0.479	−0.111 + 0.619i	−0.327 + 3.382i
7	0.570	−0.432 + 0.899i	−0.437 − 2.594i
9	0.479	0.111 − 0.619i	−0.106 + 2.144i
11	0.426	0.034 − 0.962i	−0.251 + 3.726i
13	0.580	0.026 + 3.291i	−0.459 + 2.236i
15	0.426	−0.034 + 0.962i	−0.182 + 1.801i
17	0.669	−0.341 − 1.202i	0.198 − 3.141i
19	0.603	0.475 − 1.792i	0.376 + 1.715i
21	0.640	−0.071 − 2.171i	0.351 + 3.814i
23	0.653	−0.108 + 1.382i	−0.103 − 3.800i
25	0.410	0.411 − 0.642i	0.447 − 3.701i
27	0.435	0.242 − 2.535i	0.155 + 2.315i
29	0.455	0.247 + 1.117i	0.193 − 1.213i
31	0.448	0.326 − 0.902i	−0.007 − 1.501i

reverse the orientation of $\mathfrak{U}^3$. By Theorem 1 of [4], if a glide-reflection normalizes $\pi K\theta_{12}$ then the longitude polynomial $L(t) \in Z[t]$ which annihilates the longitude entry for this equivalence class of p-reps must have even degree. In other words, the longitude entry $\lambda(\theta)$ would have to generate a field of even degree over Q. But for 9_{32} the degree $[F(\theta) : Q]$ is odd, so no subfield can have even degree. In fact, L has degree 29 and $Q(\lambda(\theta)) = F(\theta)$.

The second kind of isometry is of EH-rotations of order 2. If such a rotation were to normalize $\pi K\theta_{12}$, then 9_{32} would be invertible. As noted in [4, section 1], we have no algebraic invariant which could exclude this possibility, so we have to argue that our picture of $\mathfrak{D}$ does not have the kind of symmetry that such a rotation would require. From Table 1 we see that there are just two isometric spheres carrying

Table 2 The Edge Cycle Presentation for 9_{32}[a]

P-REP FOR 3-BRIDGE KNOT 9,32 HAS 32 NON EH-CYCLES AND 134 NON EH-EDGES.

CYCLE 1	T− 03	−A	T 03	T 01		
CYCLE 2	−A	T− 03	T 05	T 01		
CYCLE 3	T− 09	T 03	−A	T 01		
CYCLE 4	−A	T− 09	T 15	T 01		
CYCLE 5	T− 17	A	T 07	−A	T 01	
CYCLE 6	−A	T− 25	A	T 23	T 01	
CYCLE 7	−A	T− 31	T 29	T 01		
CYCLE 8	A	T− 03	T 13	T 03		
CYCLE 9	T− 05	T− 13	−A	T 03		
CYCLE 10	T− 07	−A	T− 27	T 03		
CYCLE 11	T− 09	T 13	A	T 03		
CYCLE 12	T− 07	−A	T− 21	A	T 05	
CYCLE 13	T− 15	T 13	A	T 05		
CYCLE 14	T− 15	−A	T 19	A	T 07	
CYCLE 15	T− 23	−L−A	T 21	T 07		
CYCLE 16	−A	T− 29	T− 17	T 07		
CYCLE 17	T− 11	T− 13	−A	T 09		
CYCLE 18	T− 17	T− 27	T 09			
CYCLE 19	T− 25	T 23	T 09			
CYCLE 20	T− 31	T− 19	T 09			
CYCLE 21	T− 17	T− 21	A	T 11		
CYCLE 22	T 29	A	T− 23	−L	T 11	
CYCLE 23	T− 31	T− 21	T 11			
CYCLE 24	T 21	T− 19	A	T 13		
CYCLE 25	−A	T 21	T− 27	A	T 13	
CYCLE 26	T− 23	−A	T 23	T 15		
CYCLE 27	−A	T− 19	T− 23	T 17		
CYCLE 28	−A	T 29	T− 23	−A	T 17	
CYCLE 29	−A	T− 31	T− 17	−A	T 17	
CYCLE 30	−A	T 31	T− 25	T 17		
CYCLE 31	A	T− 21	L	T 23	−A	T 19
CYCLE 32	A	T 31	T 29	T 19		

[a]Here A means $m_1\theta$, −A means $(A)^{-1}$, T n means T_n, and T − n means $(T_n)^{-1}$. The EH-relation AL = LA, L = $l_1\theta$ should be added to this presentation.

sides of $\mathfrak{D}$ of each radius 0.762, 0.570, 0.669, and seven other values. A normalizing EH-rotation R would have to permute these pairs of spheres (perhaps with help from EH-translations), and each pair imposes restrictions on the possibilities for R. These restrictions are inconsistent (use your eyes!), so R cannot exist.

The last kind is of EH-translations $z \mapsto z + c$. According to Theorem 2 of [4], if $z + c$ normalizes $\pi K\theta$ then c has the shape $(a\lambda(\theta) + b)/d$, where $a,b,d \in Z$, $a \neq 0$, and $d \geq 2$, and furthermore (because our p-reps are integral) c must be an algebraic integer. Now it happens that $\lambda(\theta)/2$ is an algebraic integer because the coefficients of its annihilating polynomial

$$L_h(t) = 2^{-29}L(2t)$$

are all rational integers, whence at least $c = \lambda(\theta_{12})/2$ is algebraically possible. But the pairs-of-spheres argument also precludes EH-translations, and 9_{32} is indeed asymmetric.

One may wonder why $\lambda(\theta)/2$ should be an algebraic integer when the EH-translation by this amount does not normalize $\pi K\theta$. For an explanation consider the reduction of the p-reps modulo two. Using Macsyma, we find that the reduction modulo two of $F_1(x)$ is irreducible in $F_2[x]$. This implies that the ideal (2) of the integers $I(\theta)$ of $F(\theta)$ is an inert prime ideal, and the p-reps of characteristic two are all equivalent. Let A, B, C be the respective images of m_1, m_2, m_3 of πK by a p-rep ϕ of characteristic two. By another Macsyma calculation we find that

$$C = A \cdot (BA)^{12}, \qquad \text{so } \pi K\phi = \langle A,B \rangle$$

Now A, B, C have order two in $\pi K\phi$, and a group generated by two elements of order two is dihedral. Because the image of a longitude of πK in a dihedral group is trivial, $\lambda(\theta)$ must belong to a prime ideal of $I(\theta)$ dividing (2). But (2) is already prime in $I(\theta)$, so $\lambda(\theta) \in (2)$, $\lambda(\theta) = 2\lambda'(\theta)$ where $\lambda'(\theta)$ is an integer of $F(\theta)$, as claimed. We can also see that $\pi K\phi$ is the dihedral group D_{59} of order 118. For the determinant of 9_{32} is 59, whence $\pi K\phi$ must be a dihedral group D_n where $n \mid 59$. (Alternatively, the classification of the dihedral subgroups of

$PSL_2(F_q)$ stated in II(iii) of Theorem C of [1] implies that $D_n \subset PSL_2(F_q)$, $q = 2^{29}$, must have $n \mid 59$.) But A, B are visibly distinct elements of $\pi K\phi$, so $n \neq 1$, as desired.

We regret not including at least the annihilating polynomial $L_h(t)$ of $\lambda(\theta)/2$, because its coefficients are so much smaller than those of the longitude polynomial itself, and $L_h(t)$ is a valuable invariant of knot type. Alas, its coefficients run to 19 digits, which is still too much. But $L_h(t)$ has one serious application: because the inverse longitude polynomial

$$R(t) = t^{29}L\left(\frac{1}{t}\right) = (2t)^{29}L_h\left(\frac{2}{t}\right)$$

does not take the values ±1 when t is a nonzero integer, 9_{32} has the parabolic property P; see Proposition 3 of [3].

REFERENCES

1. R. Riley, Parabolic representations of knot groups. I, Proc. London Math. Soc. (3) 24, (1972), 217–242.
2. R. Riley, Parabolic representations of knot groups. II, Proc. London Math. Soc. (3) 31, (1975), 495–512.
3. R. Riley, Knots with the parabolic property P, Q. J. Math. Oxford (2) 25, (1974), 273–283.
4. R. Riley, Seven excellent knots, Brown and Thickstun, Low-Dimensional Topology, Vol. I, Cambridge University Press (1982).

Index